# Modern Birkhäuser Classics

Many of the original research and survey monographs in pure and applied mathematics published by Birkhäuser in recent decades have been groundbreaking and have come to be regarded as foundational to the subject. Through the MBC Series, a select number of these modern classics, entirely uncorrected, are being re-released in paperback (and as eBooks) to ensure that these treasures remain accessible to new generations of students, scholars, and researchers.

# Tata Lectures on Theta I

David Mumford

With the collaboration of
C. Musili, M. Nori, E. Previato,
and M. Stillman

Reprint of the 1983 Edition

Birkhäuser
Boston • Basel • Berlin

David Mumford
Brown University
Division of Applied Mathematics
Providence, RI 02912
U.S.A.

Originally published as Volume 28 in the series *Progress in Mathematics*

Cover design by Alex Gerasev.

Mathematics Subject Classification (2000): 01-02, 01A60, 11-02, 14-02, 14K25, 30-02, 32-02, 33-02, 46-02 (primary); 11E45, 11G10, 14C30, 58F07 (secondary)

**Library of Congress Control Number:** 2006936982

ISBN-10: 0-8176-4572-1 e-ISBN-10: 0-8176-4577-2
ISBN-13: 978-0-8176-4572-4 e-ISBN-13: 978-0-8176-4577-9

Printed on acid-free paper.



9 8 7 6 5 4 3 2 1

*www.birkhauser.com* (IBT)

David Mumford
With the collaboration of
C. Musili, M. Nori, E. Previato,
and M. Stillman

# Tata Lectures on Theta I

Birkhäuser

Boston • Basel • Berlin

David Mumford
Department of Mathematics
Harvard University
Cambridge, MA 02138

**Library of Congress Cataloging-in-Publication Data**
Mumford, David.
Tata lectures on theta I.
(Progress in mathematics ; v. 28)
Includes bibliographical references.
Contents: 1. Introduction and motivation : theta functions in one variable ; Basic results on theta functions in several variables.
1. Functions, Theta. I. Title. II. Series: Progress in mathematics (Cambridge, Mass.) ; 28.
QA345.M85 1982 515.9'84 82-22619
ISBN 0-8176-3109-7 (Boston)
ISBN 3-7643-3109-7 (Basel)

CIP- Kurztitelaufnahme der Deutchen Bibliothek
Mumford, David:
Tata lectures on theta / David Mumford. With the assistance of C. Musili ... - Boston; Basel; Berlin : Birkhäuser.
1. Containing introduction and motivation: theta functions in one variable, basic results on theta functions in several variables. -1982.
(Progress in mathematics ; Vol. 28)
ISBN 3-7643-3109-7
NE: GT

Printed on acid-free paper.

Third Printing 1994

*Birkhäuser*



ISBN 0-8176-3109-7
ISBN 3-7643-3109-7

Printed and bound by Quinn-Woodbine, Woodbine, New Jersey.
Printed in the USA.

9 8 7 6 5 4 3

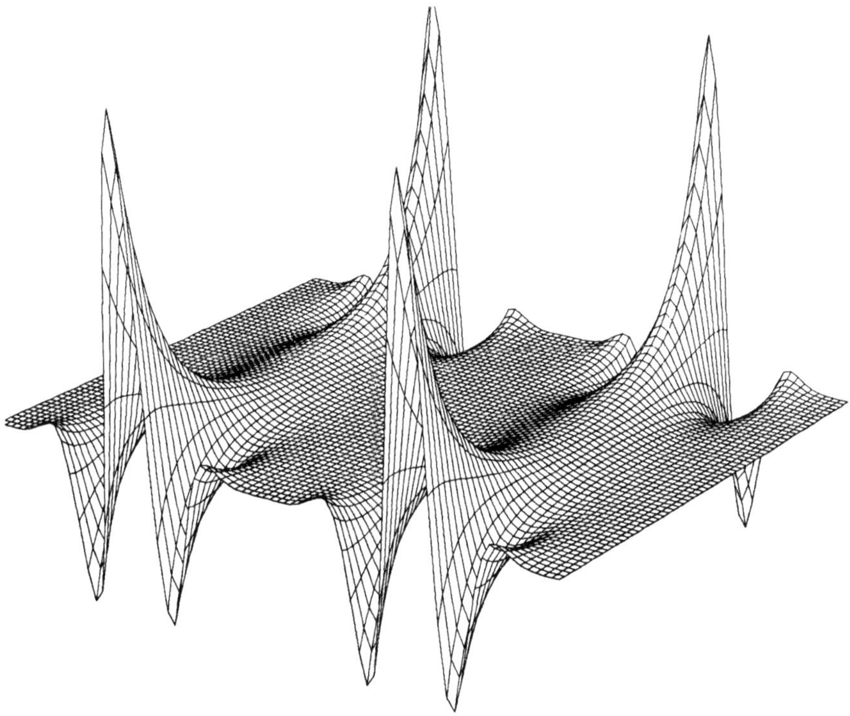

Graph of Re $\vartheta(z, \frac{i}{10})$,

$-0.5 \leq \text{Re } z \leq 1.5$

$-0.3 \leq \text{Im } z \leq 0.3$

TABLE OF CONTENTS

# Introduction

This volume contains the first two out of four chapters which are intended to survey a large part of the theory of theta functions. These notes grew out of a series of lectures given at the Tata Institute of Fundamental Research in the period October, 1978, to March, 1979, on which notes were taken and excellently written up by C. Musili and M. Nori. I subsequently lectured at greater length on the contents of Chapter III at Harvard in the fall of 1979 and at a Summer School in Montreal in August, 1980, and again notes were very capably put together by E. Previato and M. Stillman, respectively. Both the Tata Institute and the University of Montreal publish lecture note series in which I had promised to place write-ups of my lectures there. However, as the project grew, it became clear that it was better to tie all these results together, rearranging and consolidating the material, and to make them available from one place. I am very grateful to the Tata Institute and the University of Montreal for permission to do this, and to Birkhauser-Boston for publishing the final result.

The first 2 chapters study theta functions strictly from the viewpoint of classical analysis. In particular, in Chapter I, my goal was to explain in the simplest cases why the theta functions attracted attention. I look at Riemann's theta function $\vartheta(z,\tau)$ for $z \in \mathbb{C}$, $\tau \in H$ = upper half plane, also known as $\vartheta_{00}$, and its 3 variants $\vartheta_{01}, \vartheta_{10}, \vartheta_{11}$. We show how these can be used to embed the torus $\mathbb{C}/\mathbb{Z}+\mathbb{Z}\cdot\tau$ in complex projective 3-space, and how the equations for the image curve can be found. We then prove the functional equation for $\vartheta$ with respect to $SL(2,\mathbb{Z})$ and show how from this the moduli space of 1-dimensional tori itself can be realized as an algebraic curve. After this, we prove a beautiful identity of Jacobi on the z-derivative of $\vartheta$. The rest of the chapter is devoted to 3 arithmetic applications of theta series: first to some famous combinatorial identities that follow from the product expansion of $\vartheta$; second to Jacobi's formula for the number of representations of a positive integer as the sum of 4 squares; and lastly to the link between $\vartheta$ and $\zeta$ and a quick introduction to part of Hecke's theory relating modular forms and Dirichlet series.

The second chapter takes up the generalization of the geometric results of Ch. I (but not the arithmetic ones) to theta functions in several variables, i.e., to $\vartheta(\vec{z},\Omega)$ where $\vec{z} \in \mathbb{C}^g$ and $\Omega \in \mathfrak{H}_g$ = Siegel's g×g upper half-space. Again we show how $\vartheta$ can be used to embed the g-dimensional tori $X_\Omega$ in projective space. We show how, when $\Omega$ is the period matrix of a compact Riemann surface C, $\vartheta$ is related to the function theory of C. We prove the functional equation for $\vartheta$ and Riemann's theta formula, and sketch how the latter leads to explicit equations for $X_\Omega$ as an algebraic variety and to equations for certain modular schemes. Finally we show how from $\vartheta(\vec{z},\Omega)$ a large class of modular forms $\vartheta^{P,Q}(\Omega)$ can be constructed via pluri-harmonic polynomials P and quadratic forms Q.

The third chapter will study theta functions when $\Omega$ is a period matrix, i.e., Jacobian theta functions, and, in particular, hyperelliptic theta functions. We will prove an important identity of Fay from which most of the known special identities for Jacobian theta functions follow, e.g., the fact that they satisfy the non-linear differential equation known as the K-P equation. We will study at length the special properties of hyperelliptic theta functions, using an elementary model of hyperelliptic Jacobians that goes back, in its essence, to work of Jacobi himself. This leads us to a characterization of hyperelliptic period matrices $\Omega$ by the vanishing of some of the functions $\vartheta[{a \atop b}](0,\Omega)$. One of the goals is to understand hyperelliptic theta functions in their own right well enough so as to be able to deduce directly that functions derived from them satisfy the Korteweg-de Vries equation and other "integrable" non-linear differential equations.

The fourth chapter is concerned with the explanation of the group-representation theoretic meaning of theta functions and the algebro-geometric meaning of theta functions. In particular, we show how $\vartheta(\vec{z},\Omega)$ is, up to an elementary factor, a matrix coefficient of the so-called Heisenberg-Weil representation. And we show how the introduction of finite Heisenberg groups allows one to define theta functions for abelian varieties over arbitrary fields.

The third and fourth chapters will use some algebraic geometry, but the chapters in this volume assume only a knowledge of elementary classical analysis. There are several other excellent books on theta functions available and one might well ask — why another? I wished to bring out several aspects of the theory that I felt were nowhere totally clear: one is the theme that with theta functions many theories that are treated abstractly can be made very concrete and explicit, e.g., the projective embeddings, the equations for, and the moduli of complex tori. Another is the way the Heisenberg group runs through the theory as a unifying thread. However, except for the discussion in Ch. I when $g = 1$, we have not taken up the arithmetic aspects of the theory: Siegel's theory of the representation of one quadratic form by another or the Hecke operators for general g. Nor have we discussed any of the many ideas that have come recently from Shimura's idea of "lifting" modular forms. We want therefore to mention the other important books the reader may consult:

a) J. Fay, Theta Functions on Riemann Surfaces, is the best book on Jacobian theta functions (Springer Lecture Notes 352).

b) E. Freitag, Siegelsche Modulfunktionen, develops the general theory of Siegel modular forms (introducing Hecke operators and the $\phi$-operator) and the Siegel modular variety much further.

c) J.-I. Igusa, Theta functions, Springer-Verlag, 1972, like our Chapter IV unifies the group-representation theoretic and algebro-geometric viewpoints. The main result is the explicit projective embedding of the Siegel modular variety by theta constants.

d) G. Lion and M. Vergne, The Weil representation, Maslov index and Theta series (this series, No. 6) discuss the algebra of the metaplectic group on the one hand, and the theory of lifting and the Weil representation on the other. (This is the only treatment of lifting that I have been able to understand.)

The theory of theta functions is far from a finished polished topic. Each chapter finishes with a discussion of some of the unsolved problems. I hope that this book will help to attract more interest to some of these fascinating questions.

Chapter I:

# Introduction and motivation: theta functions in one variable

## § 1. Definition of $\vartheta(z,\tau)$ and its periodicity in z.

The central character in our story is the analytic function $\vartheta(z,\tau)$ in 2 variables defined by

$$\vartheta(z,\tau) = \sum_{n\in\mathbb{Z}} \exp(\pi i n^2\tau + 2\pi i n z)$$

where $z \in \mathbb{C}$ and $\tau \in H$, the upper half plane $\operatorname{Im}\tau > 0$. It is immediate that the series converges absolutely and uniformly on compact sets; in fact, if

$$|\operatorname{Im} z| < c \text{ and } \operatorname{Im}\tau > \epsilon$$

then

$$|\exp(\pi i n^2\tau + 2\pi i n z)| < (\exp -\pi\epsilon)^{n^2} . (\exp 2\pi c)^n$$

hence, if $n_o$ is chosen so that

$$(\exp -\pi\epsilon)^{n_o} . (\exp 2\pi c) < 1,$$

then the inequality

$$|\exp(\pi i n^2\tau + 2\pi i n z)| < (\exp -\pi\epsilon)^{n(n-n_o)}$$

shows that the series converges and that too very rapidly.

We may think of this series as the Fourier series for a function in z, periodic with respect to $z \longmapsto z+1$,

$$\vartheta(z,\tau) = \sum_{n\in\mathbb{Z}} a_n(\tau) \exp(2\pi i n z),\ a_n(\tau) = \exp(\pi i n^2\tau)$$

which displays the obvious fact that

$$\vartheta(z+1,\tau) = \vartheta(z,\tau)$$

The peculiar form of the Fourier coefficients is explained by examining the periodic behaviour of $\vartheta$ with respect to $z \mapsto z+\tau$: thus we have

$$\begin{aligned}\vartheta(z+\tau,\tau) &= \sum_{n\in\mathbb{Z}} \exp(\pi i n^2\tau + 2\pi i n(z+\tau)) \\ &= \sum_{n\in\mathbb{Z}} \exp(\pi i(n+1)^2\tau - \pi i\tau + 2\pi i n z) \\ &= \sum_{m\in\mathbb{Z}} \exp(\pi i m^2\tau - \pi i\tau + 2\pi i m z - 2\pi i z) \quad \text{where } m = n+1 \\ &= \exp(-\pi i\tau - 2\pi i z).\,\vartheta(z,\tau)\end{aligned}$$

so that $\vartheta$ has a kind of periodic behaviour with respect to the lattice $\Lambda_\tau \subset \mathbb{C}$ generated by 1 and $\tau$. In fact the 2 periodicities together are easily seen to give:

$$\vartheta(z+a\tau+b,\tau) = \exp(-\pi i a^2\tau - 2\pi i a z)\vartheta(z,\tau).$$

Conversely, suppose that we are looking for entire functions $f(z)$ with the simplest possible quasi-periodic behaviour w.r.t. $\Lambda_\tau$: we know by Liouville's theorem that $f$ cannot actually be periodic in 1 and $\tau$, so we may try the simplest more general possibilities:

$$f(z+1) = f(z) \text{ and } f(z+\tau) = \exp(az+b)\,.\,f(z)\,.$$

By the first, we expand $f$ in a Fourier series

$$f(z) = \sum_{n\in\mathbb{Z}} a_n \exp(2\pi i n z),\ a_n \in \mathbb{C}.$$

Writing $f(z+\tau+1)$ in terms of $f(z)$ by combining the functional equations in either order, we find that

$$f(z+\tau+1) = f(z+\tau) = \exp(az+b) \cdot f(z)$$

and also

$$f(z+\tau+1) = \exp(a(z+1)+b)\, f(z+1)$$

$$= \exp a.\ \exp(az+b)\, f(z)$$

hence $a = 2\pi ik$ for some $k \in \mathbb{Z}$. Now substituting the Foruier series into the second equation, we find that

$$\sum_{n\in\mathbb{Z}} a_n \exp(2\pi\, in\tau)\ .\ \exp(2\pi inz)$$

$$= f(z+\tau)$$

$$= \exp(2\pi ikz + b)\, f(z)$$

$$= \sum_{n\in\mathbb{Z}} a_n \exp(2\pi i(n+k)z)\ .\ \exp b$$

$$= \sum_{n\in\mathbb{Z}} a_{n-k} \exp b.\ \exp(2\pi inz).$$

Or, equivalently, for all $n \in \mathbb{Z}$, we have

(*) $$a_n = a_{n-k} \exp(b - 2\pi\, in\,\tau)\ .$$

If $k = 0$, this shows immediately that $a_n \neq 0$ for at most one $n$ and we have the uninteresting possibility that $f(z) = \exp(2\pi iz)$. If $k \neq 0$, we get a recursive relation for solving for $a_{n+kp}$ in terms of $a_n$ for all $p$. For instance, if $k = -1$, we find easily that

$$a_n = a_o \exp(-nb + \pi in(n-1)\tau) \quad \text{for all } n \in \mathbb{Z}\ .$$

This means that

$$f(z) = a_o \sum_{n\in\mathbb{Z}} \exp(-nb - \pi in\,\tau) \exp(\pi in^2\tau + 2\pi inz)$$

$$= a_o \vartheta(-z - \tfrac{1}{2}\tau - b/2\pi i\ ,\ \tau).$$

If $k > 0$, the recursion relation (*) leads to rapidly growing coefficients $a_n$ and hence there are no such entire functions $f(z)$. On the other hand, if $k < -1$, we will find a $|k|$-dimensional vector space of possibilities for $f(z)$ that will be studied in detail below. This explains the significance of $\vartheta(z, \tau)$ as an entire function of $z$ for fixed $\tau$, i.e., $\vartheta(z, \tau)$ is the most general entire function with 2 quasi-periods.

§ 2. $\vartheta(x, it)$ as the fundamental periodic solution to the Heat equation.

In a completely different vein, we may restrict the variables $z, \tau$ to the case of $z = x \in \mathbb{R}$ and $\tau = it$, $t \in \mathbb{R}^{>}$. Then

$$\vartheta(x, it) = \sum_{n \in \mathbb{Z}} \exp(-\pi n^2 t) \exp(2\pi inx)$$

$$= 1+2 \sum_{n \in \mathbb{N}} \exp(-\pi n^2 t) \cos(2\pi nx) .$$

Thus $\vartheta$ is a real valued function of 2 real variables. It satisfies the following equations:

(a) periodicity in $x$: $\vartheta(x+1, it) = \vartheta(x, it)$

(b) Heat equation:

$$\frac{\partial}{\partial t}(\vartheta(x, it)) = 2 \sum_{n \in \mathbb{N}} (-\pi n^2) \exp(-\pi n^2 t) \cos(2\pi nx)$$

$$\frac{\partial^2}{\partial x^2}(\vartheta(x, it)) = 2 \sum_{n \in \mathbb{N}} (-4\pi^2 n^2) \exp(-\pi n^2 t) \cos(2\pi nx)$$

Or

$$\frac{\partial}{\partial t}(\vartheta(x, it)) = \frac{1}{4\pi} \frac{\partial^2}{\partial x^2}(\vartheta(x, it)) .$$

This suggests that we characterise the theta function $\vartheta(x, it)$ as the unique solution to the heat equation with a certain periodic initial data when $t = 0$.

To examine the limiting behaviour of $\vartheta(x, it)$ as $t \longrightarrow 0$, we integrate it against a test periodic function

$$f(x) = \sum a_m \exp(2\pi imx).$$

Then

$$\int_0^1 \vartheta(x, it)f(x)dx = \int_0^1 \sum_{n,m} a_m \exp(-\pi n^2 t) \cdot \exp(2\pi i(n+m)x)\, dx$$

$$= \sum_{n,m} a_m \exp(-\pi n^2 t) \int_0^1 \exp(2\pi i(n+m)x)\, dx$$

$$= \sum_n a_{-n} \exp(-\pi n^2 t)$$

Therefore, we get that

$$\lim_{t \to 0} \int_0^1 \vartheta(x, it)\, f(x)\, dx$$

$$= \lim_{t \to 0} \sum_n a_n \exp(-\pi n^2 t)$$

$$= \sum_n a_n$$

$$= f(0)\ .$$

Hence $\vartheta(x, it)$ converges, as a distribution, to the sum of the delta functions at all integral points $x \in \mathbb{Z}$ as $t \longrightarrow 0$. We shall see below that it converges very nicely, in fact. Thus $\vartheta(x, it)$ may be seen as the fundamental solution to the heat equation when the space variable $x$ lies on a circle $\mathbb{R}/\mathbb{Z}$.

## § 3. The Heisenberg group and theta functions with characteristics.

In addition to the standard theta functions discussed so far, there are variants called "theta functions with characteristics" which play a very important role in understanding the functional equation and the identities satisfied by $\vartheta$, as well as the application of $\vartheta$ to elliptic curves. These are best understood group-theoretically. To explain this, let us fix a $\tau$ and then rephrase the definition of the theta function $\vartheta(z, \tau)$ by introducing transformations as follows:

For every holomorphic function $f(z)$ and real numbers $a$ and $b$, let

$$(S_b f)(z) = f(z + b)$$
$$(T_a f)(z) = \exp(\pi i a^2 \tau + 2\pi i a z) f(z + a\tau).$$

Note then that

$$S_{b_1}(S_{b_2} f) = S_{b_1+b_2} f \text{ and } T_{a_1}(T_{a_2} f) = T_{a_1+a_2} f.$$

These are the so called "1-parameter groups". However, they do not commute ! We have:

$$S_b(T_a f)(z) = (T_a f)(z+b)$$
$$= \exp(\pi i a^2 \tau + 2\pi i a(z+b)) f(z+b+a\tau)$$

and

$$T_a(S_b f)(z) = \exp(\pi i a^2 \tau + 2\pi i a z)(S_b f)(z + a\tau)$$
$$= \exp(\pi i a^2 \tau + 2\pi i a z) f(z+a\tau+b)$$

and hence

$$(*) \qquad S_b \circ T_a = \exp(2\pi i a b)\, T_a \circ S_b .$$

The group of transformations generated by the $T_a$'s and $S_b$'s is the 3-dimensional group

$$\mathcal{G} = \mathbb{C}_1^* \times \mathbb{R} \times \mathbb{R}, \quad (\mathbb{C}_1^* = \{z \in \mathbb{C} / |z| = 1\})$$

where $(\lambda, a, b) \in \mathcal{G}$ stands for the transformation:

$$(U_{(\lambda, a, b)} f)(z) = \lambda (T_a \circ S_b f)(z)$$
$$= \lambda \exp(\pi i a^2 \tau + 2\pi i a z) f(z + a\tau + b).$$

Hence the group law on $\mathcal{G}$ is given by

$$(\lambda, a, b)(\lambda', a', b') = (\lambda \lambda' \exp(2\pi i b a'), a + a', b + b')).$$

Note that

$$\underline{\text{center of } \mathcal{G}} = \mathbb{C}_1^* = \underline{\text{commutator subgroup}} [\mathcal{G}, \mathcal{G}]$$

and hence $\mathcal{G}$ is a nilpotent group.

The group $\mathcal{G}$ and its representation as above are familiar from Quantum Mechanics. Because of this connection, we will call $\mathcal{G}$ the Heisenberg group. In fact, the relation (*) is simply Weyl's integrated form of the Heisenberg commutation relations. Now recall that we have the classical theorem of Von Neumann and Stone which says that $\mathcal{G}$ has a unique irreducible unitary representation in which $(\lambda, 0, 0)$ acts by $\lambda$ (identity). In fact, this representation is the following: On our space of entire functions $f(z)$, as in §1, put the norm

$$\|f\|^2 = \int_{\mathbb{C}} \exp(-2\pi y^2/\operatorname{Im}\tau) |f(x+iy)|^2 dx\, dy.$$

Let $\mathcal{H}$ be the subspace of all $f(z)$ such that $\|f\| < \infty$. Then, it is trivial to check that $U_{\lambda, a, b}$ is unitary on $\mathcal{H}$ and it can be shown that $\mathcal{H}$ is irreducible. (In fact, the Hilbert spaces $\mathcal{H}$ and $L^2(\mathbb{R})$ are canonically isomorphic as $\mathcal{G}$-modules where $\mathcal{G}$ acts on $L^2(\mathbb{R})$ by

$$(U_{\lambda, a, b} f)(x) = \lambda \exp(2\pi i a x) f(x+b)$$

$x \in \mathbb{R}, f \in L^2(\mathbb{R})$). Thus we have in hand one of the many realisations of this canonical representation of $\mathcal{G}$. However, for the moment, this is <u>not</u> needed in our development of the theory.

To return to $\vartheta$; note that the subset

$$\Gamma = \{ (1,a,b) \in \mathcal{G} \mid a,b \in \mathbb{Z} \}$$

is a subgroup of $\mathcal{G}$. By the characterisation of $\vartheta$ in § 1, we see that, upto scalars, $\vartheta$ is the unique entire function invariant under $\Gamma$. Suppose now that $\ell$ is a positive integer; set $\ell\Gamma = \{ (1, \ell a, \ell b) \} \subseteq \Gamma$ and

$$V_\ell = \{\text{entire functions } f(z) \text{ invariant under } \ell\Gamma \}.$$

Then, we have the following:

<u>Lemma 3.1</u>. An entire function $f(z)$ is in $V_\ell$ if and only if

$$f(z) = \sum_{n \in 1/\ell \mathbb{Z}} c_n \exp(\pi i n^2 \tau + 2\pi i n z)$$

such that $c_n = c_m$ if $n-m \in \ell\mathbb{Z}$. In particular, $\dim V_\ell = \ell^2$.

<u>Proof.</u> For $a,b \in \mathbb{R}$, identify $T_a$ with $(1,a,0) \in \mathcal{G}$ and $S_b$ with $(1,0,b) \in \mathcal{G}$. If $f \in V_\ell$, then by invariance of $f$ under $S_\ell \in \ell\Gamma$, it follows that

$$f(z) = \sum_{n \in 1/\ell \mathbb{Z}} c'_n \exp(2\pi i n z).$$

On the other hand; write $c'_n = c_n \exp(\pi i n^2 \tau)$ and express the invariance of $f(z)$ under $T_\ell$, a short computation shows that $c_{n+\ell} = c_n$ for all $n \in \mathbb{Z}$, as required. (Converse is obvious).

For $m \in \mathbb{N}$, let $\mu_m \subseteq \mathbb{C}_1^*$ be the group of $m^{\text{th}}$ roots of 1. For $\ell \in \mathbb{N}$, let $\mathcal{G}_\ell$ be the finite group defined as

$$\mathcal{G}_\ell = \{(\lambda, a, b)/\lambda \in \mu_{\ell^2}\ ;\ a, b \in \tfrac{1}{\ell}\mathbb{Z}\} \pmod{\ell\Gamma}$$

$$= \mu_{\ell^2} \times (\tfrac{1}{\ell}\mathbb{Z}/\ell\mathbb{Z}) \times (\tfrac{1}{\ell}\mathbb{Z}/\ell\mathbb{Z})$$

with group law given by

$$(\lambda, a, b)(\lambda', a', b') = (\lambda\lambda' \exp(2\pi i\, ba'), a+a', b+b').$$

Now the elements $S_{1/\ell}, T_{1/\ell} \in \mathcal{G}$ commute with $\ell\Gamma$ (in view of (*)) and hence act on $V_\ell$. This goes down to an action of $\mathcal{G}_\ell$ on $V_\ell$; in fact, exactly like $\mathcal{G}$, the generators $S_{1/\ell}$ of $\mathcal{G}_\ell$ act on $V_\ell$ as follows:

$$S_{1/\ell}\Big(\sum_{n \in 1/\ell\,\mathbb{Z}} c_n \exp(\pi i n^2\tau + 2\pi i n z)\Big) = \sum_{n \in 1/\ell\,\mathbb{Z}} c_n \exp(2\pi i n/\ell).\exp(\pi i n^2\tau + 2\pi i n z)$$

and

$$T_{1/\ell}\Big(\sum_{n \in 1/\ell\,\mathbb{Z}} c_n \exp(\pi i n^2\tau + 2\pi i n z)\Big) = \sum_{n \in 1/\ell\,\mathbb{Z}} c_{n-1/\ell} \exp(\pi i n^2\tau + 2\pi i n z)$$

as is easily checked. This gives us the following:

Lemma 3.2. The finite group $\mathcal{G}_\ell$ acts irreducibly on $V_\ell$.

Proof. Let $W \subseteq V_\ell$ be a $\mathcal{G}_\ell$ - stable subspace. Take a non-zero element $f \in W$, say,

$$f(z) = \sum_{n \in 1/\ell\,\mathbb{Z}} c_n \exp(\pi i n^2\tau + 2\pi i n z), \quad c_{n_0} \neq 0.$$

Operating by powers of $S_{1/\ell}$ on $f(z)$, we find in $W$:

$$\sum_{0 \le p \le \ell^2 - 1} \exp(-2\pi i n_0 p/\ell)\,.\,(S_{p/\ell}\, f)(z)$$

$$= \sum_{n \in 1/\ell\,\mathbb{Z}} c_n \Big(\sum_p \exp(2\pi i (n - n_0) p/\ell)\, \exp(\pi i n^2\tau + 2\pi i n z)\Big)$$

$$= \ell^2 c_{n_0} \Big(\sum_{n \in n_0 + \ell\mathbb{Z}} \exp(\pi i n^2\tau + 2\pi i n z)\Big).$$

Since $c_{n_o} \neq 0$, we see that $W$ contains the function

$$\sum_{n \in n_o + \ell\mathbb{Z}} \exp(\pi i n^2 \tau + 2\pi i n z).$$

Now working with $T_{1/\ell}$ instead, we find that $W$ contains similar functions for every $n_o \in 1/_\ell\mathbb{Z}/\ell\mathbb{Z}$ and hence $W = V_\ell$.

In fact, we have also the finite analogue of Neumann-Stone theorem for $\mathcal{G}_\ell$, namely, $\mathcal{G}_\ell$ has a unique irreducible representation in which $(\lambda, 0, 0)$ acts by $\lambda.$(identity), but we do not need this at this point. For our purpose, the important point to be noted is that, because of irreducibility, the action of $\mathcal{G}_\ell$ on $V_\ell$ determines a canonical basis for $V_\ell$ and $\mathcal{G}_\ell$ acts in a fixed way. The standard basis of $V_\ell$ is given by the so called theta functions $\vartheta_{a,b}$ with rational characteristics $a, b \in 1/_\ell\,\mathbb{Z}$, defined by : for $a, b \in 1/_\ell\,\mathbb{Z}$,

$$\vartheta_{a,b} = S_b T_a \vartheta = \exp(2\pi i a b)\, T_a S_b \vartheta$$

Explicitly, we have:

$$\begin{aligned}
\vartheta_{a,b}(z, \tau) &= \exp(\pi i a^2 \tau + 2\pi i a(z+b))\, \vartheta(z + a\tau + b, \tau) \\
&= \sum_{n \in \mathbb{Z}} \exp(\pi i (a^2+n^2)\tau + 2\pi i n (z + a\tau + b) + 2\pi i a (z+b)) \\
&= \sum_{n \in \mathbb{Z}} \exp(\pi i (a+n)^2 \tau + 2\pi i (n+a)(z+b)).
\end{aligned}$$

Now we see that we have:

(0) $\vartheta_{o,o} = \vartheta$

(i) $S_{b_1}(\vartheta_{a,b}) = \vartheta_{a, b+b_1}$ for $a, b_1, b \in \frac{1}{\ell}\mathbb{Z}$

(ii) $T_{a_1}(\vartheta_{a,b}) = \exp(-2\pi i a_1 b)\, \vartheta_{a_1+a, b}$, $\forall a, a_1, b \in \frac{1}{\ell}\mathbb{Z}$

(iii) $\vartheta_{a+p,b+q} = \exp(2\pi iaq)\,\vartheta_{a,b}$, $\forall p,q \in \mathbb{Z}$, $a,b \in \frac{1}{\ell}\mathbb{Z}$.

Hence (iii) shows that $\vartheta_{a,b}$, upto a constant, depends only on $a,b \in {}^1/_\ell\,\mathbb{Z}/\mathbb{Z}$. In view of Lemma 3.1 and the Fourier expansion just given for $\vartheta_{a,b}$, it is clear that as $a,b$ run through coset representatives of $1/_\ell\,\mathbb{Z}/\mathbb{Z}$, we get a basis of $V_\ell$. Note also that except for a trivial exponential factor $\vartheta_{a,b}$ is just a translate of $\vartheta$.

## § 4. Projective emmbedding of $\mathbb{C}/\mathbb{Z}+\mathbb{Z}\tau$ by means of theta functions.

The theta functions $\vartheta_{a,b}$ defined above have a very important geometric application. Take any $\ell \geq 2$. Let $E_\tau$ be the complex torus $\mathbb{C}/\Lambda_\tau$ where $\Lambda_\tau = \mathbb{Z} + \mathbb{Z}\tau$. Let $(a_i, b_i)$ be a set of coset representatives of $(1/_\ell\,\mathbb{Z}/\mathbb{Z})^2$ in $({}^1/_\ell\,\mathbb{Z})^2$, $0 \leq i \leq \ell^2 - 1$. Write $\vartheta_i = \vartheta_{a_i, b_i}$, $0 \leq i \leq \ell^2 - 1$. For all $z \in \mathbb{C}$, consider the $\ell^2$-tuple

$$(\vartheta_0(\ell z, \tau), \ldots, \vartheta_{\ell^2-1}(\ell z, \tau))$$

modulo scalars, i.e., the homogeneous coordinates of a point in the projective space $\mathbb{P}^{\ell^2-1}_{\mathbb{C}}$. (We shall check in a minute that there is no $z, \tau$ for which they are all 0). Since

$$(\vartheta_0(z+\ell, \tau), \ldots, \vartheta_{\ell^2-1}(z+\ell, \tau)) = (\vartheta_0(z, \tau), \ldots, \vartheta_{\ell^2-1}(z, \tau))$$

and

$$(\vartheta_0(z+\ell\tau, \tau), \ldots, \vartheta_{\ell^2-1}(z, \ell\tau, \tau)) = \lambda(\vartheta_0(z, \tau), \ldots, \vartheta_{\ell^2-1}(z, \tau))$$

where $\lambda = \exp(-\pi i \ell^2 \tau - 2\pi i \ell z)$, it follows that this defines a holomorphic map

$$\varphi_\ell : E_\tau \longrightarrow \mathbb{P}^{\ell^2-1}, z \longmapsto (\ldots, \vartheta_i(\ell z, \tau), \ldots).$$

To study this map, we first prove the following:

Lemma 4.1. Every $f \in V_\ell$, $f \neq 0$, has exactly $\ell^2$ zeros (counted with multiplicities) in a fundamental domain for $\mathbb{C}/\ell\Lambda_\tau$. The zeros of $\vartheta_{a,b}$ are the points $(a+p+\frac{1}{2})\tau+(b+q+\frac{1}{2})$, $p, q \in \mathbb{Z}$. (In particular, $\vartheta_i, \vartheta_j$ $(i \neq j)$ have no common zeros and so $\varphi_\ell$ is well defined).

Proof. The first part is by the standard way of counting zeros by cantour integration: choose a parallelogram as shown missing the zeros of f:

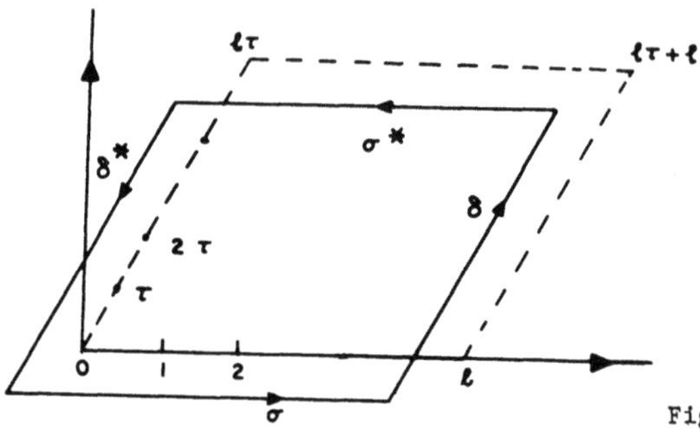


Fig. 1

Recall that we have

$$\#\text{ zeros of } f = \frac{1}{2\pi i}\int_{\sigma+\delta+\sigma^*+\delta^*}\frac{f'}{f}\,dz$$

Since $f(z+\ell) = f(z)$ and $f(z+\ell\tau) = \text{const.}\exp(-2\pi i\ell z)f(z)$, we get that

$$\int_\delta + \int_{\delta^*} = 0 \text{ and } \int_\sigma + \int_{\sigma^*} = 2\pi i\ell^2 .$$

As for the second part; note that $\vartheta(z, \tau)$ is even in z and it has a single zero in $\mathbb{C}/\Lambda_\tau$. On the other hand, we have:

$$\begin{aligned}\vartheta_{\frac{1}{2},\frac{1}{2}}(-z, \tau) &= \sum_{n\in\mathbb{Z}} \exp(\pi i(n+\tfrac{1}{2})^2\tau + 2\pi i(n+\tfrac{1}{2})(-z+\tfrac{1}{2}))\\ &= \sum_{m\in\mathbb{Z}} \exp(\pi i(-m-\tfrac{1}{2})^2\tau + 2\pi i(-m-\tfrac{1}{2})(-z+\tfrac{1}{2})) \quad \text{if } m = -1-n\\ &= \sum_{m\in\mathbb{Z}} \exp(\pi i(m+\tfrac{1}{2})^2\tau + 2\pi i(m+\tfrac{1}{2})(z+\tfrac{1}{2}) - 2\pi i(m+\tfrac{1}{2}))\\ &= -\vartheta_{\frac{1}{2},\frac{1}{2}}(z, \tau)\end{aligned}$$

and hence $\vartheta_{\frac{1}{2},\frac{1}{2}}$ is zero at $z = 0$. It follows that $\vartheta_{a,b}$ has the zeros stated and since this sum gives $\ell^2$ of them mod $\ell\Lambda_\tau$, there cannot be any more.

Next, observe that the group $\mathcal{G}_\ell$ modulo its centre, i.e., $(1/_\ell\mathbb{Z}/\ell\mathbb{Z})^2$ naturally acts on both $E_\tau$ and $\mathbb{P}^{\ell^2-1}$ and the map $\varphi_\ell$ is equivariant. To see this; let $a, b \in \frac{1}{\ell}\mathbb{Z}$, then it acts on $E_\tau$ by $z \longmapsto z + (a\tau + b)/\ell$ (and this action is free). On the other hand, if

$$U_{(1,a,b)}\,\vartheta_i = \sum_{0 \leq j \leq \ell^2-1} c_{ij}\,\vartheta_j\,,$$

the action on $\mathbb{P}^{\ell^2-1}$ is given by

$$(z_o, \ldots, z_{\ell^2-1}) \longmapsto (\sum_j c_{oj}z_j, \ldots, \sum_j c_{\ell^2-1,j}\,z_j).$$

Now we see that

$$\begin{aligned}
\varphi_\ell(z+(a\tau+b)/\ell) &= (\ldots\ldots, \vartheta_i(\ell z + a\tau + b, \tau), \ldots) \\
&= (\ldots\ldots, U_{(1,a,b)}\,\vartheta_i(\ell z, \tau), \ldots) \\
&= (\ldots\ldots, \sum_j c_{ij}\vartheta_j(\ell z, \tau)\,, \ldots)
\end{aligned}$$

and so $\varphi_\ell$ is equivariant.

It is interesting to note that, although we have a commutative group $(1/_\ell\mathbb{Z}/\ell\mathbb{Z})^2$, i.e., $(\mathbb{Z}/\ell^2\mathbb{Z})^2$, acting on $\mathbb{P}^{\ell^2-1}$, by Lemma 3.2, the action is irreducible, i.e., there are no proper invariant subspaces!

We now prove that $\varphi_\ell$ <u>is an embedding</u>: suppose, if possible, $\varphi_\ell(z_1) = \varphi_\ell(z_2)$, $z_1 \neq z_2$ in $E_\tau$, or that $d\varphi_\ell(z_1) = 0$ (the limiting case when $z_2 \longrightarrow z_1$). Translating by some $(a\tau + b)/\ell$, $a, b \in 1/_\ell\mathbb{Z}$, we find a second pair $z_1'$, $z_2'$ such that $\varphi_\ell(z_1') = \varphi_\ell(z_2')$, or $d\varphi_\ell(z_1') = 0$. Take $\ell^2-3$ further points $w_1, w_2, \ldots, w_{\ell^2-3}$, all points so far being distinct mod $\ell\Lambda_\tau$. Seek an $f \in V_\ell$, $f \neq 0$, sucn that

$$f(z_1) = f(z_1') = f(w_1) = \ldots = f(w_{\ell^2-3}) = 0.$$

This is possible because, writing $f = \sum \lambda_i \vartheta_i$, $\lambda_i \in \mathbb{C}$, we get $\ell^2-1$ linear equations in the $\ell^2$ variables $\lambda_0, \ldots, \lambda_{\ell^2-1}$ and so they have a non-zero solution. Since $\varphi_\ell(z_1) = \varphi_\ell(z_2)$, it follows that $f(z_2) = 0$. Or, if $d\varphi_\ell(z_1) = 0$, $f$ has a double zero at $z_1$. Similarly, we get that $f(z_2') = 0$ or $f$ has a double zero at $z_1'$. Therefore, $f$ has at least $\ell^2+1$ zeros in $\mathbb{C}/\ell\Lambda_\tau$, contradicting Lemma 4.1.

Thus $\varphi_\ell(E_\tau)$ is a complex analytic submanifold of $\mathbb{P}^{\ell^2-1}$ isomorphic to the torus $E_\tau$. Invoking a theorem of Chow, we can say that it is even an algebraic subvariety, i.e., $\varphi_\ell(E_\tau)$ is defined by certain homogeneous polynomials. Since everything is so explicit, we can even determine this directly. For simplicity, we will limit ourselves to the case $\ell = 2$ and show in the next section that $\varphi_2(E_\tau)$ is the subvariety of $\mathbb{P}^3$ defined by 2 quadratic equations.

## § 5. Riemann's theta relations.

Riemann's theta relation is a very basic quartic identity satisfied by $\vartheta(z, \tau)$. A whole series of such identities exist, based on any $n \times n$ integral matrix $A$ such that ${}^tAA = m^2 I_n$. Riemann's identity is based on the choice

$$A = \begin{pmatrix} 1 & 1 & 1 & 1 \\ 1 & 1 & -1 & -1 \\ 1 & -1 & 1 & -1 \\ 1 & -1 & -1 & 1 \end{pmatrix}, \quad m = 2 \text{ and } n = 4$$

The matrix identity ${}^tAA = 4I_4$ is equivalent to the identity between quadratic forms:

$$(x+y+u+v)^2 + (x+y-u-v)^2 + (x-y+u-v)^2 + (x-y-u+v)^2 = 4(x^2+y^2+u^2+v^2).$$

Fix a $\tau$ and write $\vartheta(z) = \vartheta(z, \tau)$ and $\Lambda = \Lambda_\tau$, etc. Now for all choices of $\eta \in \frac{1}{2}\Lambda/\Lambda$, we form the products $\vartheta(x+\eta)\vartheta(y+\eta)\vartheta(v+\eta)$ and sum up, putting in simple exponential factors to make the functions look like:

$$B(0): \vartheta(x)\vartheta(y)\vartheta(u)\vartheta(v) = \sum_{n,m,p,q \in \mathbb{Z}} \exp[\pi i(\Sigma n^2)\tau + 2\pi i(\Sigma xn)]$$

where $\Sigma n^2 = n^2 + m^2 + p^2 + q^2$ and $\Sigma xn = xn + ym + up + vq$

$$B(\tfrac{1}{2}): \vartheta(x+\tfrac{1}{2})\vartheta(y+\tfrac{1}{2})\vartheta(u+\tfrac{1}{2})\vartheta(v+\tfrac{1}{2})$$

$$= \sum_{n,m,p,q \in \mathbb{Z}} \exp[\pi i(\Sigma n + \Sigma n^2)\tau + 2\pi i(\Sigma xn)]$$

$$B(\tfrac{1}{2}\tau): \exp[\pi i(\tau + \Sigma x)]\vartheta(x+\tfrac{1}{2}\tau)\vartheta(y+\tfrac{1}{2}\tau)\vartheta(u+\tfrac{1}{2}\tau)\vartheta(v+\tfrac{1}{2}\tau)$$

$$= \sum_{n,m,p,q \in \mathbb{Z}} \exp[\pi i(\Sigma(n+\tfrac{1}{2})^2\tau) + 2\pi i(\Sigma x(n+\tfrac{1}{2}))]$$

$$B(\tfrac{1}{2}(1+\tau)): \exp[\pi i(\tau + \Sigma x)]\vartheta(x+\tfrac{1}{2}+\tfrac{\tau}{2})\vartheta(y+\tfrac{1}{2}+\tfrac{\tau}{2})\vartheta(u+\tfrac{1}{2}+\tfrac{\tau}{2})\vartheta(v+\tfrac{1}{2}+\tfrac{\tau}{2})$$

$$= \sum_{n,m,p,q \in \mathbb{Z}} \exp[\pi i(\Sigma n) + \pi i(\Sigma(n+\tfrac{1}{2})^2\tau) + 2\pi i(\Sigma x(n+\tfrac{1}{2}))].$$

Calling the exponential factor $e_\eta$ and summing up, we get :

$$\sum_{\eta = 0, \frac{1}{2}, \frac{1}{2}\tau, \frac{1}{2}(1+\tau)} e_\eta \vartheta(x+\eta)\vartheta(y+\eta)\vartheta(u+\eta)\vartheta(v+\eta)$$

$$= 2 \sum_{n,m,p,q \in \frac{1}{2}\mathbb{Z}} \exp[\pi i(\Sigma n^2)\tau + 2\pi i(\Sigma xn)]$$

$n, m, p, q$ <u>all</u> in $\mathbb{Z}$ or <u>all</u> in $\frac{1}{2} + \mathbb{Z}$ and $n+m+p+q \in 2\mathbb{Z}$. For simplicity, let us write

$$n_1 = \tfrac{1}{2}(n+m+p+q), \qquad x_1 = \tfrac{1}{2}(x + y+y+u+v)$$

$$m_1 = \tfrac{1}{2}(n+m-p-q), \qquad y_1 = \tfrac{1}{2}(x+y-u-v)$$

$$p_1 = \tfrac{1}{2}(n-m+p-q), \qquad u_1 = \tfrac{1}{2}(x-y+u-v)$$

$$q_1 = \tfrac{1}{2}(n-m-p+q), \qquad v_1 = \tfrac{1}{2}(x-y-u+v).$$

Note that the peculiar restrictions on the parameters $n, m, p, q$ of the summation above exactly mean that the resulting $n_1, m_1, p_1$ and $q_1$ are integers. Also observe that we have the <u>identities:</u>

$$\Sigma n^2 = \Sigma n_1^2 \quad \text{and} \quad \Sigma xn = \Sigma x_1 n_1 .$$

Now substituting these in the above equation, we get:

$$\sum_{\eta = 0, \frac{1}{2}, \frac{1}{2}\tau, \frac{1}{2}(1+\tau)} e_\eta \vartheta(x+\eta)\vartheta(y+\eta)\vartheta(u+\eta)\vartheta(v+\eta)$$

$$= 2 \sum_{n_1, m_1, p_1, q_1 \in \mathbb{Z}} \exp[\pi i(\Sigma n_1^2)\tau + 2\pi i(\Sigma x_1 n_1)].$$

Thus we have the final (Riemann's) formula (using B(0)):

$$(R_1): \quad \sum_{\eta = 0, \frac{1}{2}, \frac{1}{2}\tau, \frac{1}{2}(1+\tau)} e_\eta \vartheta(x+\eta)\vartheta(y+\eta)\vartheta(u+\eta)\vartheta(v+\eta) = 2\vartheta(x_1)\vartheta(y_1)\vartheta(u_1)\vartheta(v_1).$$

If we had started with another matrix $A$ such that ${}^tAA = m^2 I_n$, we would have found an identity of order $n$, involving summation over translates of $\vartheta$ with respect to all the division points $\eta$ of order $m$, i.e., $\eta \in \frac{1}{m}\Lambda/\Lambda$. To use the identity $(R_1)$, it is natural to reformulate it with theta functions $\vartheta_{a,b}$ with characteristics $a, b \in \frac{1}{2}\mathbb{Z}$; there are 4 of these, namely:

$$\vartheta_{0,0}(z,\tau) = \sum_{n\in\mathbb{Z}} \exp(\pi i n^2 \tau + 2\pi i n z) = \vartheta(z,\tau)$$

$$\vartheta_{0,\frac{1}{2}}(z,\tau) = \sum \exp(\pi i n^2 \tau + 2\pi i n(z+\tfrac{1}{2})) = \vartheta(z+\tfrac{1}{2},\tau)$$

$$\vartheta_{\frac{1}{2},0}(z,\tau) = \sum \exp(\pi i (n+\tfrac{1}{2})^2 \tau + 2\pi i(n+\tfrac{1}{2}) z) = \exp(\pi i \tau/4 + \pi i z)\vartheta(z+\tfrac{1}{2}\tau,\tau)$$

$$\vartheta_{\frac{1}{2},\frac{1}{2}}(z,\tau) = \sum \exp(\pi i(n+\tfrac{1}{2})^2 \tau + 2\pi i(n+\tfrac{1}{2})(z+\tfrac{1}{2})) = \exp(\pi i \tau/4 + \pi i(z+\tfrac{1}{2}))\vartheta(z+\tfrac{1}{2}(1+\tau),\tau)$$

For simplicity, we write these as $\vartheta_{00}$, $\vartheta_{01}$, $\vartheta_{10}$ and $\vartheta_{11}$. It is immediately verified that

$$\vartheta_{00}(-z,\tau) = \vartheta_{00}(z,\tau)$$

$$\vartheta_{01}(-z,\tau) = \vartheta_{01}(z,\tau)$$

$$\vartheta_{10}(-z,\tau) = \vartheta_{10}(z,\tau)$$

$$\vartheta_{11}(-z,\tau) = -\vartheta_{11}(z,\tau)$$

showing that $\vartheta_{11}$ is different from the others, and confirming the fact that $\vartheta_{11}(0,\tau) = 0$, while the other 3 are <u>not</u> zero at $z = 0$ (cf. Lemma 4.1). Now Riemann's formula gives us:

$$\begin{aligned}(R_2):\ & \vartheta_{00}(x)\vartheta_{00}(y)\vartheta_{00}(u)\vartheta_{00}(v) + \vartheta_{01}(x)\vartheta_{01}(y)\vartheta_{01}(u)\vartheta_{01}(v) \\ & + \vartheta_{10}(x)\vartheta_{10}(y)\vartheta_{10}(u)\vartheta_{10}(v) + \vartheta_{11}(x)\vartheta_{11}(y)\vartheta_{11}(u)\vartheta_{11}(v) \\ & = 2\,\vartheta_{00}(x_1)\vartheta_{00}(y_1)\vartheta_{00}(u_1)\vartheta_{00}(v_1)\end{aligned}$$

where $x_1 = \frac{1}{2}(x+y+u+v)$, $y_1 = \frac{1}{2}(x+y-u-v)$, etc. Now replacing $x$ by $x+1$ and using the fact that his changes the sign of $\vartheta_{10}$ and $\vartheta_{11}$, we get further:

$$
\begin{aligned}
(R_3): \quad & \vartheta_{00}(x)\vartheta_{00}(y)\vartheta_{00}(u)\vartheta_{00}(v) + \vartheta_{01}(x)\vartheta_{01}(y)\vartheta_{01}(u)\vartheta_{01}(v) \\
& - \vartheta_{10}(x)\vartheta_{10}(y)\vartheta_{10}(u)\vartheta_{10}(v) - \vartheta_{11}(x)\vartheta_{11}(y)\vartheta_{11}(u)\vartheta_{11}(v) \\
& = 2\,\vartheta_{01}(x_1)\vartheta_{01}(y_1)\vartheta_{01}(u_1)\vartheta_{01}(v_1).
\end{aligned}
$$

Substituting instead $x+\tau$ for $x$ in $(R_2)$ and multiplying by $\exp(\pi i\tau + 2\pi ix)$ so that $\vartheta_{00}(x)$ becomes $\vartheta_{00}(x)$ again while $\vartheta_{01}(x)$ and $\vartheta_{11}(x)$ change signs, we get:

$$
\begin{aligned}
(R_4): \quad & \vartheta_{00}(x)\vartheta_{00}(y)\vartheta_{00}(u)\vartheta_{00}(v) - \vartheta_{01}(x)\vartheta_{01}(y)\vartheta_{01}(u)\vartheta_{01}(v) \\
& + \vartheta_{10}(x)\vartheta_{10}(y)\vartheta_{10}(u)\vartheta_{10}(v) - \vartheta_{11}(x)\vartheta_{11}(y)\vartheta_{11}(u)\vartheta_{11}(v) \\
& = 2\,\vartheta_{10}(x_1)\vartheta_{10}(y_1)\vartheta_{10}(u_1)\vartheta_{11}(v_1).
\end{aligned}
$$

Finally replacing $x$ by $x+\tau+1$ in $(R_2)$ and multiplying by $\exp(\pi i\tau + 2\pi ix)$, we get:

$$
\begin{aligned}
(R_5): \quad & \vartheta_{00}(x)\vartheta_{00}(y)\vartheta_{00}(u)\vartheta_{00}(v) - \vartheta_{01}(x)\vartheta_{01}(y)\vartheta_{01}(u)\vartheta_{01}(v) \\
& - \vartheta_{10}(x)\vartheta_{10}(y)\vartheta_{10}(u)\vartheta_{10}(v) + \vartheta_{11}(x)\vartheta_{11}(y)\vartheta_{11}(u)\vartheta_{11}(v) \\
& = 2\,\vartheta_{11}(x_1)\vartheta_{11}(y_1)\vartheta_{11}(u_1)\vartheta_{11}(v_1).
\end{aligned}
$$

In other words, we get all 4 theta functions on the right hand side by putting a character into the sum on the left hand side. More variants can be obtained by similar small substitutions: viz., replacing $x, y, u$ and $v$ by $x+\alpha, y+\beta, u+\gamma, v+\delta$ where $\alpha, \beta, \gamma, \delta \in \frac{1}{2}\Lambda$ and $\alpha+\beta+\gamma+\delta \in \Lambda$. We have listed below in an abbreviated form all the results.

First we make a table containing the fundamental transformation relations between the $\vartheta_{ij}$'s that are needed for a quick verification of Riemann's theta formulae.

Table 0

| $(z \longmapsto -z)$ | $(z \longmapsto z+\frac{1}{2})$ |
|---|---|
| $\vartheta_{00}(-z,\tau) = \vartheta_{00}(z,\tau)$ | $\vartheta_{00}(z+\frac{1}{2},\tau) = \vartheta_{01}(z,\tau)$ |
| $\vartheta_{01}(-z,\tau) = \vartheta_{01}(z,\tau)$ | $\vartheta_{01}(z+\frac{1}{2},\tau) = \vartheta_{00}(z,\tau)$ |
| $\vartheta_{10}(-z,\tau) = \vartheta_{10}(z,\tau)$ | $\vartheta_{10}(z+\frac{1}{2},\tau) = \vartheta_{11}(z,\tau)$ |
| $\vartheta_{11}(-z,\tau) = -\vartheta_{11}(z,\tau)$ | $\vartheta_{11}(z+\frac{1}{2},\tau) = -\vartheta_{10}(z,\tau)$ |

$(z \longmapsto z+\frac{1}{2}\tau)$

$$\vartheta_{00}(z+\tfrac{1}{2}\tau,\tau) = (\exp(-\pi i\tau/4 - \pi i z))\,\vartheta_{10}(z,\tau)$$
$$\vartheta_{01}(z+\tfrac{1}{2}\tau,\tau) = -i(\qquad '' \qquad)\,\vartheta_{11}(z,\tau)$$
$$\vartheta_{10}(z+\tfrac{1}{2}\tau,\tau) = (\qquad '' \qquad)\,\vartheta_{00}(z,\tau)$$
$$\vartheta_{11}(z+\tfrac{1}{2}\tau,\tau) = -i(\qquad '' \qquad)\,\vartheta_{01}(z,\tau)$$

$(z \longmapsto z+\frac{1}{2}\tau+\frac{1}{2})$

$$\vartheta_{00}(z+\tfrac{1}{2}\tau+\tfrac{1}{2},\tau) = -i\,(\exp(-\pi i\tau/4 - \pi i z))\,\vartheta_{11}(z,\tau)$$
$$\vartheta_{01}(\qquad '' \qquad) = (\qquad '' \qquad)\,\vartheta_{10}(z,\tau)$$
$$\vartheta_{10}(\qquad '' \qquad) = -i(\qquad '' \qquad)\,\vartheta_{01}(z,\tau)$$
$$\vartheta_{11}(\qquad '' \qquad) = -(\qquad '' \qquad)\,\vartheta_{00}(z,\tau)$$

## RIEMANN'S THETA FORMULAE

I. $(R_1)$: $$\sum_{\eta=0,\frac{1}{2},\frac{\tau}{2},\frac{1+\tau}{2}} e_\eta \vartheta(x+\eta)\vartheta(y+\eta)\vartheta(u+\eta)\vartheta(v+\eta) = 2\vartheta(x_1)\vartheta(y_1)\vartheta(u_1)\vartheta(v_1)$$

where $e_\eta = 1$ for $\eta = 0, \frac{1}{2}$ and $e_\eta = \exp(\pi i \tau + \pi i(x+y+u+v))$ for $\eta = \frac{1}{2}(1+\tau)$, and

$x_1 = \frac{1}{2}(x+y+u+v)$, $y_1 = \frac{1}{2}(x+y-u-v)$, $u_1 = \frac{1}{2}(x-y+u-v)$ and $v_1 = \frac{1}{2}(x-y-u+v)$.

II. Via Half-integer thetas:

$\vartheta_{00}^{x} = \vartheta(x,\tau) = \sum \exp(\pi i n^2\tau + 2\pi i n x)$, $\vartheta_{01}^{x} = \sum \exp(\pi i n^2\tau + 2\pi i n(x+\frac{1}{2}))$,

$\vartheta_{10}^{x} = \sum \exp(\pi i(n+\frac{1}{2})^2\tau + 2\pi i(n+\frac{1}{2})x)$ and $\vartheta_{11}^{x} = \sum \exp(\pi i(n+\frac{1}{2})^2\tau + 2\pi i(n+\frac{1}{2})(x+\frac{1}{2}))$

$(R_2)$: $\vartheta_{00}^{x}\vartheta_{00}^{y}\vartheta_{00}^{u}\vartheta_{00}^{v} + \vartheta_{01}^{x}\vartheta_{01}^{y}\vartheta_{01}^{u}\vartheta_{01}^{v} + \vartheta_{10}^{x}\vartheta_{10}^{y}\vartheta_{10}^{u}\vartheta_{10}^{v} + \vartheta_{11}^{x}\vartheta_{11}^{y}\vartheta_{11}^{u}\vartheta_{11}^{v} = 2\vartheta_{00}^{x_i}\vartheta_{00}^{y_1}\vartheta_{00}^{u_1}\vartheta_{00}^{v_1}$

" + " − " − " $= 2\vartheta_{01}^{x_1}\vartheta_{01}^{y_1}\vartheta_{01}^{u_1}\vartheta_{01}^{v_1}$

" − " + " − " $= 2\vartheta_{10}^{x_1}\vartheta_{10}^{y_1}\vartheta_{10}^{u_1}\vartheta_{10}^{v_1}$

" − " − " + " $= 2\vartheta_{11}^{z_1}\vartheta_{11}^{y_1}\vartheta_{11}^{u_1}\vartheta_{11}^{v_1}$

$(R_6)$: $\vartheta_{00}^{x}\vartheta_{00}^{y}\vartheta_{01}^{u}\vartheta_{01}^{v} + \vartheta_{01}^{x}\vartheta_{01}^{y}\vartheta_{00}^{u}\vartheta_{00}^{v} + \vartheta_{10}^{x}\vartheta_{10}^{y}\vartheta_{11}^{u}\vartheta_{11}^{v} + \vartheta_{11}^{x}\vartheta_{11}^{y}\vartheta_{10}^{u}\vartheta_{10}^{v} = 2\vartheta_{01}^{z_{11}}\vartheta_{01}^{y_1}\vartheta_{00}^{u_1}\vartheta_{00}^{v_1}$

" + " − " − " $= 2\vartheta_{00}^{x_1}\vartheta_{00}^{y_1}\vartheta_{01}^{u_1}\vartheta_{01}^{v_1}$

" − " + " − " $= -2\vartheta_{11}^{x_1}\vartheta_{11}^{y_1}\vartheta_{10}^{u_1}\vartheta_{10}^{v_1}$

" − " − " + " $= -2\vartheta_{10}^{x_1}\vartheta_{10}^{y_1}\vartheta_{11}^{u_1}\vartheta_{11}^{v_1}$

$(R_{10})$: $\vartheta_{00}^{x}\vartheta_{00}^{y}\vartheta_{10}^{u}\vartheta_{10}^{v} + \vartheta_{01}^{x}\vartheta_{01}^{y}\vartheta_{11}^{u}\vartheta_{11}^{v} + \vartheta_{10}^{x}\vartheta_{10}^{y}\vartheta_{00}^{u}\vartheta_{00}^{v} + \vartheta_{11}^{x}\vartheta_{11}^{y}\vartheta_{01}^{u}\vartheta_{01}^{v} = 2\vartheta_{00}^{x_1}\vartheta_{00}^{y_1}\vartheta_{10}^{u_1}\vartheta_{10}^{v_1}$

" + " − " − " $= 2\vartheta_{01}^{x_1}\vartheta_{01}^{y_1}\vartheta_{11}^{u_1}\vartheta_{11}^{v_1}$

" − " + " − " $= 2\vartheta_{10}^{x_1}\vartheta_{10}^{y_1}\vartheta_{00}^{u_1}\vartheta_{00}^{v_1}$

" − " − " + " $= 2\vartheta_{11}^{x_1}\vartheta_{11}^{y_1}\vartheta_{01}^{u_1}\vartheta_{01}^{v_1}$

$(R_{14})$: $\vartheta_{00}^{x}\vartheta_{00}^{y}\vartheta_{11}^{u}\vartheta_{11}^{v} + \vartheta_{01}^{x}\vartheta_{01}^{y}\vartheta_{10}^{u}\vartheta_{10}^{v} + \vartheta_{10}^{x}\vartheta_{10}^{y}\vartheta_{01}^{u}\vartheta_{01}^{v} + \vartheta_{11}^{x}\vartheta_{11}^{y}\vartheta_{00}^{u}\vartheta_{00}^{v} = 2\vartheta_{01}^{x_1}\vartheta_{01}^{y_1}\vartheta_{10}^{u_1}\vartheta_{10}^{v_1}$

" + " − " − " $= 2\vartheta_{00}^{x_1}\vartheta_{00}^{y_1}\vartheta_{11}^{u_1}\vartheta_{11}^{v_1}$

" − " + " − " $= -2\vartheta_{11}^{x_1}\vartheta_{11}^{y_1}\vartheta_{00}^{u_1}\vartheta_{00}^{v_1}$

" − " − " + " $= -2\vartheta_{10}^{x_1}\vartheta_{10}^{y_1}\vartheta_{01}^{u_1}\vartheta_{01}^{v_1}$

$(R_{18})$: $\vartheta_{00}^{x}\vartheta_{01}^{y}\vartheta_{10}^{u}\vartheta_{11}^{v} + \vartheta_{01}^{x}\vartheta_{00}^{y}\vartheta_{11}^{u}\vartheta_{10}^{v} + \vartheta_{10}^{x}\vartheta_{11}^{y}\vartheta_{00}^{u}\vartheta_{01}^{v} + \vartheta_{11}^{x}\vartheta_{10}^{y}\vartheta_{01}^{u}\vartheta_{00}^{v} = 2\vartheta_{11}^{x_1}\vartheta_{10}^{y_1}\vartheta_{01}^{u_1}\vartheta_{00}^{v_1}$

" + " − " − " $= -2\vartheta_{10}^{x_1}\vartheta_{11}^{y_1}\vartheta_{00}^{u_1}\vartheta_{01}^{v_1}$

" − " + " − " $= -2\vartheta_{01}^{x_1}\vartheta_{00}^{y_1}\vartheta_{11}^{u_1}\vartheta_{10}^{v_1}$

$(R_{21})$: " − " − " + " $= 2\vartheta_{00}^{x_1}\vartheta_{01}^{y_1}\vartheta_{10}^{u_1}\vartheta_{11}^{v_1}$

We have listed these at such length to illustrate a key point in the theory of theta functions: the symmetry of the situation generates rapidly an overwhelming number of formulae, which do not hawever make a completely elementary pattern. To obtain a clear picture of the algebraic implications of these formulae altogether is then not usually easy.

One important consequence of these formulae comes from specialising the variables, setting $x = y$ and $u = v$. The important fact to remember is that $\vartheta_{11}(0) = 0$ whereas $\vartheta_{00}, \vartheta_{01}$ and $\vartheta_{10}$ are not zero at 0 (cf. Lemma 4.1)! Then the right hand side of $(R_5)$ is 0; and $(R_5)$, $(R_2) + (R_5)$ combine to give (noting that $x_1 = x+u, y_1 = x-u$ and $u_1 = v_1 = 0$):

$$(A_1) : \vartheta_{00}(x)^2 \vartheta_{00}(u)^2 + \vartheta_{11}(x)^2 \vartheta_{11}(u)^2 = \vartheta_{01}(x)^2 \vartheta_{01}(u)^2 + \vartheta_{10}(x)^2 \vartheta_{10}(u)^2$$

$$= \vartheta_{00}(x+u)\vartheta_{00}(x-u)\vartheta_{00}(0)^2.$$

Likewise, $(R_3) + (R_5)$ and $(R_4) + (R_5)$ (with $x = y, u = v$) respectively give:

$$(A_2) : \vartheta_{01}(x+u)\vartheta_{01}(x-u)\vartheta_{01}(0)^2 = \vartheta_{00}(x)^2\vartheta_{00}(u)^2 - \vartheta_{10}(x)^2\vartheta_{10}(u)^2$$

$$= \vartheta_{01}(x)^2 \vartheta_{01}(u)^2 - \vartheta_{11}(x)^2 \vartheta_{11}(u)^2$$

and

$$(A_3) : \vartheta_{10}(x+u)\vartheta_{10}(x-u)\vartheta_{10}(0)^2 = \vartheta_{00}(x)^2 \vartheta_{00}(u)^2 - \vartheta_{01}(x)^2 \vartheta_{01}(y)^2$$

$$= \vartheta_{10}(x)^2 \vartheta_{10}(u)^2 - \vartheta_{11}(x)^2 \vartheta_{11}(u)^2.$$

These are trypical of the "addition formulae" for theta functions for calculating the coordinates of $\varphi_2(x+u)$, $\varphi_2(x-u)$ in terms of those of $\varphi_2(x)$, $\varphi_2(u)$ and $\varphi_2(0)$. There are 12 more expressions for the products $\vartheta_{ab}(x+u)\vartheta_{cd}(x-u)$ in terms of $\vartheta_{ef}(x)$'s, $\vartheta_{ef}(u)$'s, for $ab \neq cd$, which we have written down on the next page. All are obtained from the formulae $(R_n)$ by just setting the variables equal in pairs.

## III. Addition Formulae

$$(A_1) : \vartheta_{00}(x+u)\,\vartheta_{00}(x-u)\,\vartheta_{00}^2(0) = \vartheta_{00}^2(x)\vartheta_{00}^2(u) + \vartheta_{11}^2(x)\vartheta_{11}^2(u) = \vartheta_{01}^2(x)\vartheta_{01}^2(u) + \vartheta_{10}^2(x)\vartheta_{10}^2(u)$$

$$\vartheta_{01}(x+u)\,\vartheta_{01}(x-u)\,\vartheta_{01}^2(0) = \vartheta_{00}^2(x)\vartheta_{00}^2(u) - \vartheta_{10}^2(x)\,\vartheta_{10}(u) = \vartheta_{01}^2(x)\vartheta_{01}^2(u) - \vartheta_{11}^2(x)\vartheta_{11}^2(u)$$

$$\vartheta_{10}(x+u)\,\vartheta_{10}(x-u)\,\vartheta_{10}^2(0) = \vartheta_{00}^2(x)\,\vartheta_{00}^2(u) - \vartheta_{01}^2(x)\vartheta_{01}^2(u) = \vartheta_{10}^2(x)\vartheta_{10}^2(u) - \vartheta_{11}^2(x)\vartheta_{11}^2(u)$$

$$\vartheta_{00}(x+u)\vartheta_{01}(x-u)\vartheta_{00}(0)\vartheta_{01}(0) = \vartheta_{00}(x)\vartheta_{01}(x)\vartheta_{00}(u)\vartheta_{01}(u) - \vartheta_{01}(x)\vartheta_{11}(x)\vartheta_{01}(u)\vartheta_{11}(u)$$

$$\vartheta_{01}(x+u)\vartheta_{00}(x-u)\vartheta_{00}(0)\vartheta_{01}(0) = \vartheta_{00}(x)\vartheta_{01}(x)\vartheta_{00}(u)\vartheta_{01}(u) + \vartheta_{01}(x)\vartheta_{11}(x)\vartheta_{01}(u)\vartheta_{11}(u)$$

$$\vartheta_{00}(x+u)\vartheta_{10}(x-u)\vartheta_{00}(0)\vartheta_{01}(0) = \vartheta_{00}(x)\vartheta_{10}(x)\vartheta_{00}(u)\vartheta_{10}(u) + \vartheta_{01}(x)\vartheta_{11}(x)\vartheta_{01}(u)\vartheta_{11}(u)$$

$$\vartheta_{10}(x+u)\vartheta_{00}(x-u)\vartheta_{00}(0)\vartheta_{10}(0) = \vartheta_{00}(x)\vartheta_{10}(x)\vartheta_{00}(u)\vartheta_{10}(u) - \vartheta_{01}(x)\vartheta_{11}(x)\vartheta_{01}(u)\vartheta_{11}(u)$$

$$\vartheta_{01}(x+u)\vartheta_{10}(x-u)\vartheta_{01}(0)\vartheta_{10}(0) = \vartheta_{00}(x)\vartheta_{11}(x)\vartheta_{00}(u)\vartheta_{11}(u) + \vartheta_{01}(x)\vartheta_{10}(x)\vartheta_{01}(u)\vartheta_{10}(u)$$

$$\vartheta_{10}(x+u)\vartheta_{01}(x-u)\vartheta_{01}(0)\vartheta_{10}(0) = -\vartheta_{00}(x)\vartheta_{11}(x)\vartheta_{00}(u)\vartheta_{11}(u) + \vartheta_{01}(x)\vartheta_{10}(x)\vartheta_{01}(u)\vartheta_{10}(u)$$

$$(A_{10}) : \vartheta_{11}(x+u)\vartheta_{11}(x-u)\vartheta_{00}^2(0) = \vartheta_{11}^2(x)\vartheta_{00}^2(u) - \vartheta_{00}^2(x)\vartheta_{11}^2(u) = \vartheta_{01}^2(x)\vartheta_{10}^2(u) - \vartheta_{10}^2(x)\vartheta_{01}^2(u)$$

$$\vartheta_{11}(x+u)\vartheta_{00}(x-u)\vartheta_{01}(0)\,\vartheta_{10}(0) = \vartheta_{00}(x)\vartheta_{11}(x)\vartheta_{01}(u)\vartheta_{10}(u) + \vartheta_{10}(x)\vartheta_{01}(x)\vartheta_{00}(u)\vartheta_{11}(u)$$

$$\vartheta_{00}(x+u)\vartheta_{u}(x-u)\,\vartheta_{01}(0)\,\vartheta_{10}(0) = \vartheta_{00}(x)\,\vartheta_{11}(x)\vartheta_{01}(u)\vartheta_{10}(u) - \vartheta_{10}(x)\vartheta_{01}(x)\vartheta_{00}(u)\vartheta_{11}(u)$$

$$\vartheta_{11}(x+u)\vartheta_{01}(x-u)\vartheta_{00}(0)\vartheta_{10}(0) = \vartheta_{00}(x)\vartheta_{10}(x)\vartheta_{01}(u)\vartheta_{11}(u) + \vartheta_{01}(x)\vartheta_{11}(x)\vartheta_{00}(u)\vartheta_{10}(u)$$

$$\vartheta_{01}(x+u)\vartheta_{11}(x-u)\vartheta_{00}(0)\vartheta_{10}(0) = -\vartheta_{00}(x)\vartheta_{10}(x)\vartheta_{01}(u)\vartheta_{11}(u) + \vartheta_{01}(x)\vartheta_{11}(x)\vartheta_{00}(u)\vartheta_{10}(u)$$

$$\vartheta_{11}(x+u)\vartheta_{10}(x-u)\vartheta_{00}(0)\vartheta_{01}(0) = \vartheta_{00}(x)\vartheta_{01}(x)\vartheta_{10}(u)\vartheta_{11}(u) + \vartheta_{10}(x)\vartheta_{11}(x)\vartheta_{00}(u)\vartheta_{01}(u)$$

$$\vartheta_{10}(x+u)\vartheta_{11}(x-u)\vartheta_{00}(0)\vartheta_{01}(0) = -\vartheta_{00}(x)\vartheta_{01}(x)\vartheta_{10}(u)\vartheta_{11}(u) + \vartheta_{10}(x)\vartheta_{11}(x)\vartheta_{00}(u)\vartheta_{01}(u)$$

## IV. Equations for $\vartheta$

$$(E_1) : \vartheta_{00}^2(x)\,\vartheta_{00}^2(0) = \vartheta_{01}^2(x)\,\vartheta_{01}^2(0) + \vartheta_{10}^2(x)\,\vartheta_{10}^2(0)$$

$$(E_2) : \vartheta_{11}^2(x)\,\vartheta_{00}^2(0) = \vartheta_{01}^2(x)\,\vartheta_{10}^2(0) - \vartheta_{10}(x)\,\vartheta_{01}^2(x) \quad \text{and}$$

$$(J_1) : \vartheta_{00}^4(0) = \vartheta_{01}^4(0) + \vartheta_{10}^4(0)$$

Specialising further by setting $u = 0$, we find that all the above reduce to just 2 relations:

$$(E_1): \qquad \vartheta_{00}(x)^2\,\vartheta_{00}(0)^2 = \vartheta_{01}(x)^2\,\vartheta_{01}(0)^2 + \vartheta_{10}(x)^2\,\vartheta_{10}(0)^2$$

$$(E_2): \qquad \vartheta_{11}(x)^2\,\vartheta_{00}(0)^2 = \vartheta_{01}(x)^2\,\vartheta_{10}(0)^2 - \vartheta_{10}(x)^2\,\vartheta_{01}(0)^2 .$$

Finally setting $x = 0$, we obtain Jacobi's identity between the "theta constants" $\vartheta_{ab}^{(0)}$, namely:

$$(M_1) \qquad \vartheta_{00}(0)^4 = \vartheta_{01}(0)^4 + \vartheta_{10}(0)^4 .$$

We see now that the identities $(E_1)$ and $(E_2)$ are equations satisfied by the image $\varphi_2(E_\tau)$ in $\mathbb{P}^3$. We now appeal to some simple algebraic geometry to conclude that $\varphi_2(E_\tau)$ is indeed the curve $C$ in $\mathbb{P}^3$ defined by the following 2 quadratic equations:

$$\vartheta_{00}(0)^2 x_0^2 = \vartheta_{01}(0)^2\, x_1^2 + \vartheta_{10}(0)^2\, x_2^2$$

$$\vartheta_{00}(0)^2 x_3^2 = \vartheta_{10}(0)^2 x_1^2 - \vartheta_{01}(0)^2 x_2^2$$

By Bezout's theorem, it is clear that a hyperplane $H$ in $\mathbb{P}^3$ meets $C$ in utmost 4 points. But the hyperplane $\Sigma a_i x_i = 0$ meets $\varphi_2(E_\tau)$ in the points where

$$a_0\vartheta_{00}(2x) + a_1\vartheta_{01}(2x) + a_2\vartheta_{10}(2x) + a_3\vartheta_{11}(2x) = 0,$$

and there are 4 such points mod $2\Lambda_\tau$. So $\varphi_2(E_\tau)$ must be equal to $C$! It is clear that the theta relations give us explicit formulae for everything that goes on in the curve $C$.

## § 6. Doubly periodic meromorphic functions via $\vartheta(z, \tau)$.

By means of theta functions, there are 4 ways of defining meromorphic functions on the elliptic curve $E_\tau$, and the above identities enable us to relate them:

Method I: By restriction of rational functions from $\mathbb{P}^3$: This gives the basic meromorphic functions

$$\frac{\vartheta_{ab}(2z)}{\vartheta_{oo}(2z)}$$

on $E_\tau$.

Method II. As quotients of products of translates of $\vartheta(z)$ itself: i.e., if $a_1, \ldots, a_k, b_1, \ldots, b_k \in \mathbb{C}$ are such that $\Sigma a_i = \Sigma b_i$, then it is easy to check that

$$\prod_{1 \leq i \leq k} \frac{\vartheta(z-a_i)}{\vartheta(z-b_i)}$$

is periodic for $\Lambda_\tau$, hence is a meromorphic function with zeros at $a_i + \frac{1}{2}(1+\tau)$ and poles at $b_i + \frac{1}{2}(1+\tau)$. (If we use $\vartheta_{11}$ instead of $\vartheta$, we get zeros at $a_i$ and poles at $b_i$). In fact, all meromorphic functions arise like this and this expression is just like the prime factorisation of meromorphic functions on $\mathbb{P}^1$:

$$f(z) = \prod_{1 \leq i \leq k} \frac{(a_i z_1 - b_i z_o)}{(c_i z_1 - d_i z_o)}$$

$z = z_1/z_o$, $z_1, z_o$ homogeneous coordinates, zeros at $z = b_i/a_i$ and poles at $z = d_i/c_i$.

Method III. Second logarithmic derivatives: Note that $\log \vartheta(z)$ is periodic upto addition of a linear function. Thus the (doubly) periodic function

$$\frac{d^2}{dz^2} \log \vartheta(z)$$

is meromorphic. This is essentially Weierestrass' $\wp$ -function. To be precise,

$$\wp(z) = -\frac{d^2}{dz^2} \log \vartheta_{11}(z) + (\text{constant}),$$

the constant being adjusted so that the Laurent expansion of $\wp(z)$ at $z = 0$ has no constant term.

Method IV: Sums of first logarithmic derivatives: Choose $a_1, \dots, a_k \in \mathbb{C}$ and $\lambda_1, \dots, \lambda_k \in \mathbb{C}$ such that $\sum \lambda_i = 0$. Then one checks that

$$\sum_i \lambda_i \frac{d}{dz} \log \vartheta(z - a_i) + (\text{constant})$$

is periodic for $\Lambda_\tau$, hence is meromorphic with simple poles at $a_i + \frac{1}{2}(1+\tau)$, residues $\lambda_i$. Again, this gives all meromorphic functions with simple poles and is the analogue of the partial fraction expansion for meromorphic functions on $\mathbb{P}^1$:

$$f(z) = \sum_i \frac{\lambda_i}{(z - a_i)} + (\text{constant})$$

We give a few of the relations between these functions: for example, to relate Methods I and II, we need merely expand each $\vartheta_{ab}(2z)$ as a product of 4 functions $\vartheta(z-a_i)$, times an exponential factor (the $a_i$'s being the zeros of $\vartheta_{ab}(2z)$). For instance, $\vartheta_{11}(2z)$ is 0 at the 4 (2-torsion) points $0, \frac{1}{2}, \frac{1}{2}\tau, \frac{1}{2}(1+\tau)$, and its factorisation:

$$\vartheta_{11}(2z)\,\vartheta_{00}(0)\,\vartheta_{01}(0)\,\vartheta_{10}(0) = 2\,\vartheta_{00}(z)\,\vartheta_{01}(z)\,\vartheta_{10}(z)\,\vartheta_{11}(z)$$

is the formula $(R_{18})$ when $x = y = u = v = z$. To relate Methods II and III, we can take the 2nd derivative with respect to $u$ in the formula $(A_{10})$ and set $x = z$ and $u = 0$. We get:

$$(\vartheta_{11}(z)\vartheta_{11}''(z) - \vartheta_{11}'(z)^2)\,\vartheta_{00}(0)^2 = \vartheta_{11}(z)^2\,\vartheta_{00}(0)\,\vartheta_{00}''(0) - \vartheta_{00}(z)^2\,\vartheta_{11}'(0)^2$$

(using the fact that $\vartheta_{01}'(0) = 0$ and $\vartheta_{11}''(0) = 0$ since $\vartheta_{01}$ is an even function and $\vartheta_{11}$ is an odd function). Dividing both sides by $\vartheta_{00}(0)^2\vartheta_{11}(z)^2$, we find that the resulting equation is simply

$$\frac{d^2}{dz^2}\log\vartheta_{11}(z) = \frac{\vartheta_{00}''(0)}{\vartheta_{00}(0)} - \frac{\vartheta_{11}'(0)^2}{\vartheta_{00}(0)^2}\cdot\frac{\vartheta_{00}(z)^2}{\vartheta_{11}(z)^2}$$

hence

$$\wp(z) = (\text{constant}) + \frac{\vartheta_{11}'(0)^2}{\vartheta_{00}(0)^2}\cdot\frac{\vartheta_{00}(z)^2}{\vartheta_{11}(z)^2}\ .$$

One of the most important facts about the $\wp$-function is the differential equation that it satisfies. In fact, using the obvious facts (from the above equation) that (i) $\wp(z) = \wp(-z)$, (ii) the expansion of $\wp(z)$ at $z = 0$ begins with $z^{-2}$ and (iii) the constant rigged so that this expansion has no constant term, it follows that:

$$\wp(z) = \frac{1}{z^2} + az^2 + bz^4 + \ldots, \text{ near } z = 0.$$

Therefore,

$$\wp'(z) = -\frac{2}{z^3} + 2az + 4bz^3 + \ldots$$

and hence

$$(\wp'(z))^2 = \frac{4}{z^6} - \frac{8a}{z^2} - 16b + \ldots$$

But

$$4\wp(z)^3 = \frac{4}{z^6} + \frac{12a}{z^2} + 12b + \ldots$$

so that

$$\wp'(z)^2 - 4\wp(z)^3 + 20\,a\,\wp(z) = -28b + \ldots$$

Thus the function $\wp'(z)^2 - 4\wp(z)^3 + 20a\wp(z)$ is a doubly periodic entire function and hence is a constant. This means we have an identity:

$$\wp'(z)^2 = 4\wp(z)^3 + g_2(\tau)\wp(z) + g_3(\tau)$$

which is Weierstrass' differential equation for $\wp(z)$. Differentiating this twice, we also get the differential equations:

$$\wp''(z) = 6\wp(z)^2 + g_2(\tau)$$

and

$$\wp'''(z) = 12\wp(z)\,.\wp'(z)$$

Thus $\wp$ is a time independent solution of the (Kortweg-de Vries) KdV (non-linear wave) equation:

$$u_t = u_{xxx} - 12uu_x\,, \quad u = u(x,t)$$

§ 7. The functional equation of $\vartheta(z, \tau)$.

So far we have concentrated on the behaviour of $\vartheta(z, \tau)$ as a function of $z$. Its behaviour as a function of $\tau$ is also extremely beautiful, but rather deeper and more subtle. Just as $\vartheta$ is periodic upto an elementary factor for a group of transformations acting in $z$, so also it is periodic upto a factor for a group acting on $z$ and $\tau$. To derive this so called "functional equation" of $\vartheta$ in $\tau$; note that if we consider $\vartheta(z,\tau)$ for a fixed $\tau$, then although its definition involves the generators 1 and $\tau$ of the lattice $\Lambda_\tau$ quite unsymmetrically, still in its application to $E_\tau$ (describing the function theory and projective embedding of $E_\tau$) this asymmetry disappears. In other words, if we had picked any 2 other generators $a\tau+b, c\tau+d$ of $\Lambda_\tau$ $(a,b,c,d \in \mathbb{Z},\ ad-bc = \pm 1)$, we could have constructed theta functions which were periodic with respect to $z \longmapsto z+c\tau+d$ and periodic upto an exponential factor for $z \longmapsto z+a\tau+b$, and these theta functions would be equally useful for the study of $E_\tau$. Clearly then however the new theta functions could not be too different from the original ones! If we try to make this connection precise, we are lead immediately to the functional equation of $\vartheta(z, \tau)$ in $\tau$.

To be precise, fix any

$$\begin{pmatrix} a & b \\ c & d \end{pmatrix} \in SL(2, \mathbb{Z}), \text{ i.e. } a,b,c,d \in \mathbb{Z},\ ad-bc = +1$$

and assume that $ab, cd$ are even. Multiplying by $-1$ if necessary, we assume that $c \geq 0$.

Consider the function $\vartheta((c\tau+d)\,y, \tau)$. Clearly, when $y$ is replaced by $y+1$, the function is unchanged except for an exponential factor. It is not hard to rig up an exponential factor of the type $\exp(Ay^2)$ which corrects $\vartheta((c\tau+d)y, \tau)$ to a periodic function for $y \mapsto y+1$. In fact, let

$$\Upsilon(y,\tau) = \exp(\pi ic(c\tau+d)y^2)\,\vartheta((c\tau+d)y,\tau).$$

Then a simple calculation shows that

$$\Upsilon(y+1,\tau) = \Upsilon(y,\tau)$$

(N.B: a factor $\exp(\pi icd)$ appears, so we use $cd$ even in the verification). However, the periodic behaviour of $\vartheta$ for $z \mapsto z+\tau$ gives a 2nd quasi-period for $\Upsilon$, namely,

$$\Upsilon\left(y+\frac{a\tau+b}{c\tau+d},\tau\right) = \exp\left(-\pi i\frac{a\tau+b}{c\tau+d} - 2\pi iy\right)\Upsilon(y,\tau).$$

We give some of the calculations this time: formally writing we have by definition:

$$\frac{\Upsilon\left(y+\frac{a\tau+b}{c\tau+d},\tau\right)}{\vartheta((c\tau+d)y+a\tau+b,\tau)} = \exp\left[\pi ic(c\tau+d)y^2 + 2\pi icy(a\tau+b) + \pi ic\frac{(a\tau+b)^2}{c\tau+d}\right].$$

But

$$\frac{\vartheta((c\tau+d)y+a\tau+b,\tau)}{\Upsilon(y,\tau)} = \frac{\exp[-\pi ia^2\tau - 2\pi iay(c\tau+d)]\vartheta((c\tau+d)y,\tau)}{\exp(\pi ic(c\tau+d)y^2)\,\vartheta((c\tau+d)y,\tau)}$$

$$= \exp(-\pi ia^2\tau - 2\pi iay(c\tau+d) - \pi ic(c\tau+d)y^2).$$

So multiplying these two equations and using $ad-bc = 1$, we get:

$$\frac{\Upsilon\left(y+\frac{a\tau+b}{c\tau+d},\tau\right)}{\Upsilon(y,\tau)} = \exp\left(-2\pi iy(ad-bc) + \pi ic\frac{(a\tau+b)^2}{(c\tau+d)} - \pi ia^2\tau\right)$$

$$= \exp\left[-2\pi iy - \frac{\pi i}{c\tau+d}(a^2\tau(c\tau+d) - c(a\tau+b)^2)\right]$$

$$= \exp\left[-\pi iy - \frac{\pi i}{c\tau+d}(a^2\tau d - 2abc\tau - b^2c)\right]$$

But

$$a^2\tau d - 2abc\tau - b^2c = a(ad-bc)\tau - ab(c\tau+d) + b(ad-bc)$$

$$= (a\tau+b) - ab(c\tau+d).$$

Now using ab is even, we get what we want. If we now recall the characterisation of $\vartheta(y, \tau')$ as a function of $y$ as in § 1, namely, $\vartheta(y, \tau')$ is the unique function (upto scalars) invariant under $\Lambda_{\tau'}$ where $\tau' = (a\tau+b)/(c\tau+d)$), we find $\Upsilon(y, \tau)$ is one such. Hence we get:

$$\Upsilon(y, \tau) = \varphi(\tau)\, \vartheta(y, (a\tau+b)/(c\tau+d))$$

for some function $\varphi(\tau)$. In other words, if $y = z/(c\tau+d)$, then

$$\vartheta(z, \tau) = \varphi(\tau) \exp(-\pi i c z^2/(c\tau+d))\, \vartheta(z/(c\tau+d), (a\tau+b)/(c\tau+d)).$$

To evaluate $\varphi(\tau)$; note that $\vartheta(z, \tau)$ is normalised by the property that the $0^{th}$ term in its Fourier expansion is just 1, i.e.,

$$\int_0^1 \vartheta(y, \tau)\, dy = 1.$$

Hence

$$\varphi(\tau) = \int_0^1 \Upsilon(y, \tau)\, dy = \int_0^1 \exp(\pi i c(c\tau+d) y^2)\, \vartheta((c\tau+d) y, \tau)\, dy .$$

This integral is fortunately not too hard to calculate. First note that $\varphi(\tau) = d (= \pm 1)$ if $c = 0$ and so we can assume $c > 0$. Now substituting the defining series for $\vartheta$ and rearranging terms, we get:

$$\varphi(\tau) = \int_0^1 \sum_{n\in\mathbb{Z}} \exp[\pi i (cy+n)^2(\tau+d/c) - \pi i n^2 d/c]\, dy$$

$$= \sum_{n\in\mathbb{Z}} \exp(-\pi i n^2 d/c) \int_0^1 \exp(\pi i (cy+n)^2(\tau+d/c))\, dy.$$

But (using again cd even) we have

$$\exp(-\pi i d(n+c)^2/c) = \exp(-\pi i n^2 d/c)$$

and hence we get

$$\varphi(\tau) = \sum_{1\leq n\leq c} \exp(-\pi i n^2 d/c) \int_{-\infty}^{\infty} \exp \pi i c^2 y^2(\tau + d/c)\, dy .$$

To evaluate the integral; first suppose that $\tau = it - d/c$. Then we get

$$\int_{-\infty}^{\infty} \exp \pi i c^2 y^2(\tau+d/c)\, dy = \int_{-\infty}^{\infty} \exp(-\pi c^2 y^2 t)\, dy$$

i.e., if $u = c\, t^{\frac{1}{2}} y$,

$$= \frac{1}{c\, t^{\frac{1}{2}}} \int_{-\infty}^{\infty} \exp(-\pi u^2)\, du$$

$$= 1/c\, t^{\frac{1}{2}} ,$$

using the well-known fact that $\int_{-\infty}^{\infty} \exp(-\pi u^2)\, du = 1$. It then follows by analytic continuation that for any $\tau$ with $\operatorname{Im} \tau > 0$, we have

$$\int_{-\infty}^{\infty} \exp \pi i c^2 y^2 (\tau + d/c) dy = \frac{1}{c[(\tau+d/c)/i]^{\frac{1}{2}}}$$

where $(\ \ )^{\frac{1}{2}}$ is chosen such that $\operatorname{Re}(\ \ )^{\frac{1}{2}} > 0$. The sum is a well-known "Gauss sum"

$$S_{d,c} = \sum_{1\leq n\leq c} \exp(-\pi i n^2 d/c)$$

which in fact is just $c^{\frac{1}{2}}$ times some $8^{th}$ root of 1. This may be proved either directly (but not too easily) from number theory, or it can be deduced, by induction on $c + |d|$, from the <u>compatibilities</u> of the functional equations. In fact, the exact functional equation is given in the following:

Theorem 7.1. Given $a, b, c, d \in \mathbb{Z}$ such that $ad-bc = 1$, $ab$ and $cd$ even, we have then for a suitable $\zeta$, an $8^{th}$ root of 1, that

$$(F_1): \quad \vartheta\left(\frac{z}{c\tau+d}, \frac{a\tau+b}{c\tau+d}\right) = \zeta(c\tau+d)^{\frac{1}{2}} \exp(\pi icz^2/(c\tau+d))\, \vartheta(z, \tau).$$

To fix $\zeta$ exactly, we consider two cases: first assume $c > 0$ or $c = 0$ and $d > 0$ (multiplying $\begin{pmatrix} a & b \\ c & d \end{pmatrix}$ by $-1$ if necessary), hence $\mathrm{Im}(c\tau+d) \geq 0$ and choose $(c\tau+d)^{\frac{1}{2}}$ in the first quadrant ($\mathrm{Re}(\ ) \geq 0$ and $\mathrm{Im}(\ ) \geq 0$):

(a) if $c$ is even and $d$ is odd, then

$$\zeta = i^{\frac{1}{2}(d-1)}\left(\frac{c}{|d|}\right)$$

where $\left(\frac{x}{y}\right)$ is the Jacobi symbol (to take care of all cases, we set $\left(\frac{0}{1}\right) = +1$),

(b) if $c$ is odd and $d$ is even, then

$$\zeta = \exp(-\pi ic/4)\left(\frac{d}{c}\right).$$

Proof. We have only to see (a) and (b); we first check it for two special cases (i) and (ii):

(i) $$\begin{pmatrix} a & b \\ c & d \end{pmatrix} = \begin{pmatrix} 1 & b \\ 0 & 1 \end{pmatrix}, \quad b \text{ even}.$$

Now $(F_1)$ + (a) says the obvious identity, namely,

$$(F_2): \qquad \vartheta(z, \tau+b) = \vartheta(z, \tau)$$

(ii) $$\begin{pmatrix} a & b \\ c & d \end{pmatrix} = \begin{pmatrix} 0 & -1 \\ 1 & 0 \end{pmatrix}.$$

Now $(F_1)$ + (b) reads as

$$(F_3): \qquad \vartheta(z/\tau, -1/\tau) = \exp(-\pi i/4).\ \tau^{\frac{1}{2}}.\ \exp(\pi iz^2/\tau)\,\vartheta(z, \tau)$$

which we have already proved since it is trivial that $S_{0,1} = 1$. We get the general case by induction on $|c| + |d|$: if $|d| > |c|$, we substitute $\tau \pm 2$ for

$\tau$ in $(F_1)$ and use $(F_2)$ to show that $(F_1)$ for a, b, c, d follows from $(F_1)$ for $a, b \pm 2a, c, d \pm 2c$. Since we can make $|d \pm 2c| < |d|$, we are done. Note that $|d \pm 2c| \neq |d|$ or $|c|$ because $(c,d) = 1$ and cd is even. On the other hand, if $|d| < |c|$, we substitute $-1/\tau$ for $\tau$ in $(F_1)$ and use $(F_3)$ to show that $(F_1)$ for a, b, c, d follows from $(F_1)$ for b, -a, d, -c: this reduces us to the case $|d| > |c|$ again. The details are lengthy (and hence omitted) but straight forward (the usual properties of the Jacobi symbol, e.g., reciprocity, must be used). It is, however, à priori clear that the method must give a function equation of type $(F_1)$ for some $8^{th}$ root $\zeta$ of 1.

§ 8. The Heat equation again.

The transformation formula for $\vartheta(z,\tau)$ allows us to see very explicitly what happens to the real valued function $\vartheta(x, it)$, studied in § 2, when $t \longrightarrow 0$. In fact, $(F_3)$ says:

$$\vartheta(x/it, i/t) = t^{\frac{1}{2}} \exp(\pi x^2/t)\, \vartheta(x, it)$$

hence

$$\begin{aligned}\vartheta(x, it) &= t^{-\frac{1}{2}} \exp(-\pi x^2/t) \sum_{n \in \mathbb{Z}} \exp(-\pi n^2/t + (2\pi nx/t)) \\ &= t^{-\frac{1}{2}} \sum_{n \in \mathbb{Z}} \exp(-\pi (x-n)^2/t).\end{aligned}$$

In completely elementary terms, this is the rather striking identity:

$$1 + 2 \sum_{n \in \mathbb{N}} \cos(2\pi nx) \exp(-\pi n^2 t) = t^{-\frac{1}{2}} \sum_{m \in \mathbb{Z}} \exp(-\pi (x-m)^2/t).$$

But $t^{-\frac{1}{2}} \exp(-\pi x^2/t)$ is the well-known fundamental solution to the Heat equation on the line, with initial data at $t = 0$ being a delta function at $x = 0$. Thus $\vartheta(x, it)$ is just the superposition of infinitely many such solutions, with initial data being delta functions at integer values $x = n$. In particular, this shows that $\vartheta(x, it)$ is positive and goes to 0 as $t \longrightarrow 0$ uniformly when $1-x \geq x \geq \epsilon$.

§ 9. The concept of modular forms.

Let us stand back from our calculation now and consider what we have got so far. In the first place, the substitutions in the variables $z, \tau$ for which $\vartheta$ is quasi-periodic form a group: in fact, $SL(2, \mathbb{Z})$ acts on $\mathbb{C} \times H$ by

$$(z, \tau) \longmapsto (z/c\tau+d, (a\tau+b)/(c\tau+d))$$

because

$$\left(\frac{z/(c\tau+d)}{c'((a\tau+b)/(c\tau+d))+d'}, \frac{a'((a\tau+b)/(c\tau+d))+b'}{c'((a\tau+b)/(c\tau+d))+d'}\right)$$

$$= \left(\frac{z}{(c'a+d'c)\tau+(c'b+dd')}, \frac{(a'a+b'c)\tau+(a'b+b'd)}{(c'a+d'c)\tau+(c'b+d'd)}\right)$$

Moreover, this action normalises the lattice action on $z$, i.e., we have an action of a semi-direct product

$$SL(2, \mathbb{Z}) \ltimes \mathbb{Z}^2$$

on $\mathbb{C} \times H$, where $(\begin{pmatrix} a & b \\ c & d \end{pmatrix}; m, n)$ acts by

$$(z, \tau) \longmapsto ((z+m\tau+n)/(c\tau+d), (a\tau+b)/(c\tau+d)).$$

Actually; not all of these carry $\vartheta$ to itself; we put on the side condition $ab, cd$ even. To understand this condition group theoretically; note that we have a natural homomorphism

$$\gamma_N : SL(2, \mathbb{Z}) \longrightarrow SL(2, \mathbb{Z}/N\mathbb{Z})$$

for every N. Its kernel $\Gamma_N$, the so called "level N-principal congruence subgroup" is given by

$$\Gamma_N = \{ \begin{pmatrix} a & b \\ c & d \end{pmatrix} \in SL(2,\mathbb{Z}) / b, c \equiv 0 \pmod N,\ a, d \equiv 1 \pmod N \}.$$

Before we study the level 2 case explicitly, let us recall that the group $SL(2,\mathbb{Z}/2\mathbb{Z})$ of six matrices

$$\begin{pmatrix} 1 & 0 \\ 0 & 1 \end{pmatrix}, \begin{pmatrix} 0 & 1 \\ 1 & 0 \end{pmatrix}, \begin{pmatrix} 1 & 1 \\ 0 & 1 \end{pmatrix}, \begin{pmatrix} 1 & 1 \\ 1 & 0 \end{pmatrix}, \begin{pmatrix} 1 & 0 \\ 1 & 1 \end{pmatrix}, \begin{pmatrix} 0 & 1 \\ 1 & 1 \end{pmatrix}$$

is isomorphic to the group of permutation on 3 letters (1, 2, 3):

$$(1)(2)(3), (12)(3), (23)(1), (123), (13)(2), (132).$$

We define, following Igusa, $\Gamma_{1,2} \subset SL(2,\mathbb{Z})$ to be $\gamma_2^{-1}$ of the subgroup of $SL(2,\mathbb{Z}/2\mathbb{Z})$ consisting of $\begin{pmatrix} 1 & 0 \\ 0 & 1 \end{pmatrix}$ and $\begin{pmatrix} 0 & 1 \\ 1 & 0 \end{pmatrix}$. Clearly this is the subset of $SL(2,\mathbb{Z})$ of elements $\begin{pmatrix} a & b \\ c & d \end{pmatrix}$ such that ab and cd are even. Note however that whereas $\Gamma_N$ is a normal subgroup of $SL(2,\mathbb{Z})$, $\Gamma_{1,2}$ is <u>not</u>; it has 2 conjugates:

$$\gamma_2^{-1}\left(\begin{pmatrix} 1 & 0 \\ 0 & 1 \end{pmatrix}, \begin{pmatrix} 1 & 1 \\ 0 & 1 \end{pmatrix}\right) \text{ and } \gamma_2^{-1}\left(\begin{pmatrix} 1 & 0 \\ 0 & 1 \end{pmatrix}, \begin{pmatrix} 1 & 0 \\ 1 & 1 \end{pmatrix}\right)$$

described by the conditions c even and b even respectively. They are the groups for which $\vartheta_{01}$ and $\vartheta_{10}$ have functional equations. If we write out

$$\vartheta(z/(c\tau+d), (a\tau+b)/(c\tau+d))$$

when $\begin{pmatrix} a & b \\ c & d \end{pmatrix} \notin \Gamma_{1,2}$, we find that it is an elementary factor times $\vartheta_{01}(z,\tau)$ or $\vartheta_{10}(z,\tau)$. The simplest way to see this is not to try to describe how an arbitrary $\begin{pmatrix} a & b \\ c & d \end{pmatrix}$ in $SL(2,\mathbb{Z})$ transforms the $\vartheta_{ij}$'s - which leads to interminable problems of sign - but rather to consider the action of 2 generators

$\begin{pmatrix} 1 & 1 \\ 0 & 1 \end{pmatrix}$ and $\begin{pmatrix} 0 & -1 \\ 1 & 0 \end{pmatrix}$ of SL(2, $\mathbb{Z}$). Their action is summarised in the following table:

Table V

| | |
|---|---|
| $\vartheta_{00}(z, \tau+1) = \vartheta_{01}(z, \tau)$ | $\vartheta_{00}(z/\tau, -1/\tau) = (-i\tau)^{\frac{1}{2}} \exp(\pi i z^2/\tau)\ \vartheta_{00}(z, \tau)$ |
| $\vartheta_{01}(\ ''\ ) = \vartheta_{00}(\ ''\ )$ | $\vartheta_{01}(\ ''\ ) = \ ''\ \vartheta_{10}(\ ''\ )$ |
| $\vartheta_{10}(\ ''\ ) = \exp(\pi i/4)\vartheta_{10}('')$ | $\vartheta_{10}(\ ''\ ) = \ ''\ \vartheta_{01}(\ ''\ )$ |
| $\vartheta_{11}(\ ''\ ) = \exp(\pi i/4)\vartheta_{11}('')$ | $\vartheta_{11}(\ ''\ ) = -(\ ''\ )\ \vartheta_{11}(\ ''\ )$ |

From this, the action of any $\begin{pmatrix} a & b \\ c & d \end{pmatrix}$ can be described. (The formulae on the left are verified directly by substitution in the Fourier expansion. The 1st formula on the right is $(F_3)$. The 2nd comes for instance by substituting $z + \frac{1}{2}\tau$ for $z$ in the 1st and using the functional equation of $\vartheta_{00}$ in $z$; the 3rd comes from substitutions $z/\tau$ for $z$, $-\tau^{-1}$ for $\tau$ in the 2nd ; and the 4th from substituting $z+\frac{1}{2}\tau$ for $z$ in the 3rd).

Geometrically, the reason the funny subgroup $\Gamma_{1,2}$ arises is that $\vartheta(z, \tau)$ is 0 at the special point of order 2, namely, $\frac{1}{2}(\tau+1) \in \frac{1}{2}\Lambda_\tau/\Lambda_\tau$, and it is easy to check that $\begin{pmatrix} a & b \\ c & d \end{pmatrix} \in \Gamma_{1,2}$ if and only if $z \longmapsto z/(c\tau+d)$ carries $\frac{1}{2}(\tau+1)$ to $\frac{1}{2}(\tau'+1)$ mod $\Lambda_{\tau'}$, where $\tau' = (a\tau+b)/(c\tau+d)$.

However, we shall focus our attention in this section on the behaviour of the functions $\vartheta_{ij}(0, \tau)$ of <u>one</u> variable $\tau$. Note then that the functional equation of $\vartheta(0, \tau)$ reduces to:

$$\vartheta(0, (a\tau+b)/(c\tau+d)) = \zeta(c\tau+d)^{\frac{1}{2}}\vartheta(0, \tau)$$

where $\zeta^8 = 1$, $\zeta$ as given in Theorem 7.1. This will show that $\vartheta(0, \tau)^2$ is a modular form in $\tau$ in the following sense:

Definition 9.1. Let $k \in \mathbb{Z}^+$ & $N \in \mathbb{N}$. By a modular form of weight $k$ & level $N$, we mean a holomorphic function $f(\tau)$ on the upper half-plane $H$ such that

(a) for all $\tau \in H$ & $\begin{pmatrix} a & b \\ c & d \end{pmatrix} \in \Gamma_N$,

$$f((a\tau+b)/(c\tau+d)) = (c\tau+d)^k f(\tau)$$

(b) $f$ is bounded as follows:

(i) $\exists$ constants $c$ & $d$ such that $|f(\tau)| \leq c$ if $\mathrm{Im}\,\tau > d$ and

(ii) $\forall\, p/q \in \mathbb{Q}$, $\exists$ positive reals $c_{p,q}$ & $d_{p,q}$ such that

$$|f(\tau)| \leq c_{p,q} |\tau - (p/q)|^{-k} \quad \text{if } |\tau - p/q - id_{p,q}| < d_{p,q}\,.$$

The set of modular forms of weight $k$ & level $N$ is a vector space and is denoted by $\mathrm{Mod}_k^{(N)}$. Thus any $f \in \mathrm{Mod}_k^{(N)}$ is bounded outside an horizontal strip; and the circles of radii $d_{p,q}$ centred at $p/q + id_{p,q}$ (i.e., touching the real axis at the rational points $p/q$) are called the horocircles for $f$.

See
Fig. 2

Note that $SL(2, \mathbb{Z})$ acts on the "rational boundary points" $\mathbb{Q} \cup \{\infty\}$ of $H$ and that if

$$f((a\tau+b)/(c\tau+d)) = (c\tau+d)^k f(\tau)\,,$$

then the bound at $p/q \in \mathbb{Q} \cup \{\infty\}$ is equivalent to the bound at $\dfrac{a(p/q)+b}{c(p/q)+d}$ (the bound at $\infty$ being the condition (b) (i) : $|f(\tau)| \leq c$ if $\mathrm{Im}\,\tau > d$).

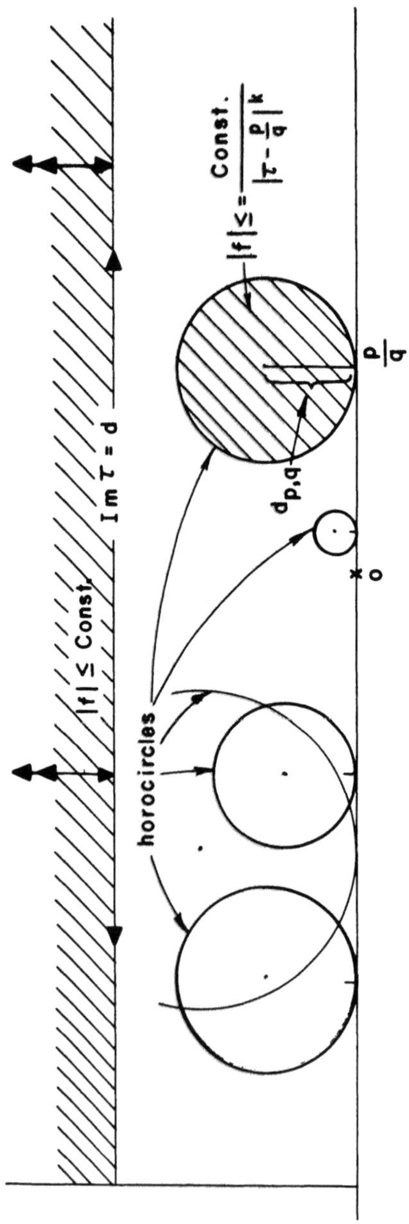


Fig. 2

The condition that makes this definition work is that the factors $(c\tau+d)^k$ introduced in the functional equation (a) satisfy the "1-cocycle" condition, i.e., if we write

$$e_\gamma(\tau) = (c\tau+d)^k, \text{ where } \gamma = \begin{pmatrix} a & b \\ c & d \end{pmatrix},$$

then for all $\gamma_1, \gamma_2 \in \Gamma_N$, we have

$$e_{\gamma_1\gamma_2}(\tau) = e_{\gamma_1}(\gamma_2\tau)\, e_{\gamma_2}(\tau).$$

This same condition, together with the fact that $\Gamma_N$ is normal in $SL(2,\mathbb{Z})$, gives an action of $SL(2,\mathbb{Z})/\Gamma_N$ on the vector space $Mod_k^{(N)}$ : if $f$ is a modular form and $\gamma = \begin{pmatrix} a & b \\ c & d \end{pmatrix}$, define

$$f^\gamma(\tau) = e_\gamma(\tau)^{-1} f(\gamma\tau).$$

It is immediate that this is also a modular form of the same kind as $f$ : in fact, (a) for $f$ implies (a) for $f^\gamma$ and (b) for $f$ at $p/q$ implies (b) for $f^\gamma$ at $\gamma(p/q)$. Finally note that if $f \in Mod_k^{(N)}$ and $g \in Mod_\ell^{(N)}$, then the product $fg \in Mod_{k+\ell}^{(N)}$. Thus

$$Mod^{(N)} = \bigoplus_{k\in\mathbb{Z}^+} Mod_k^{(N)}$$

is a graded ring, called the <u>ring of modular forms of level</u> N. Now we have the following:

<u>Proposition 9.2.</u> $\vartheta_{00}^2(0,\tau)$, $\vartheta_{01}^2(0,\tau)$ and $\vartheta_{10}^2(0,\tau)$ are modular forms of weight 1 & level 4.

<u>Proof.</u> To start with, condition (a) for $\vartheta_{00}^2(0,\tau)$ amounts to saying that $\zeta$, the 8th root of 1, in the functional equation $(F_1)$ is $\pm 1$ when $\begin{pmatrix} a & b \\ c & d \end{pmatrix} \in \Gamma_4$. This is immediate from the description of $\zeta$ (in fact, we only need $c$ even and $d \equiv 1 \pmod 4$). We can also verify immediately the bound (b)(d) at $\infty$ for

$\vartheta_{00}^2(0,\tau)$. In fact, the Fourier expansion

$$\vartheta_{00}(0,\tau) = \sum_{n \in \mathbb{Z}} \exp(\pi i n^2 \tau)$$

shows that, as $\operatorname{Im} \tau \longrightarrow \infty$, we have

$$\vartheta_{00}(0,\tau) = 1 + O(\exp(-\pi \operatorname{Im} \tau))$$

hence $\vartheta_{00}^2(0,\tau)$ is every close to 1 when $\operatorname{Im} \tau >> 0$. Before verifying (b)(ii) at the finite cusps $p/q \in \mathbb{Q}$, consider how $SL(2,\mathbb{Z})$ acts on $\vartheta_{00}^2(0,\tau)$. Let $\alpha = \begin{pmatrix} 1 & 1 \\ 0 & 1 \end{pmatrix}$ & $\beta = \begin{pmatrix} 0 & -1 \\ 1 & 0 \end{pmatrix}$ be the generators of $SL(2,\mathbb{Z})$. Using Table V above, we check the following:

$$[\vartheta_{00}^2(0,\tau)]^\alpha = \vartheta_{01}^2(0,\tau), \quad [\vartheta_{00}^2(0,\tau)]^\beta = -i\vartheta_{00}^2(0,\tau)$$

$$[\vartheta_{01}^2(0,\tau)]^\alpha = \vartheta_{00}^2(0,\tau), \quad [\vartheta_{01}^2(0,\tau)]^\beta = -i\,\vartheta_{10}^2(0,\tau)$$

$$[\vartheta_{10}^2(0,\tau)]^\alpha = i\vartheta_{10}^2(0,\tau), \quad [\vartheta_{10}^2(0,\tau)]^\beta = -i\vartheta_{01}^2(0,\tau)\,.$$

So these three give an $SL(2,\mathbb{Z})$ -invariant subspace of $\mathrm{Mod}_1^{(4)}$. But then to check the bound at the finite cusps, it suffices to check it for all 3 functions at $\infty$ because a suitable $\gamma \in SL(2,\mathbb{Z})$ carries any cusp to $\infty$. As for $\vartheta_{00}(0,\tau)$, by the Fourier expansions, we have:

$$\left.\begin{aligned} \vartheta_{01}(0,\tau) &= 1+O(\exp(-\pi \operatorname{Im} \tau)) \\ \vartheta_{10}(0,\tau) &= O(\exp(-\pi \operatorname{Im} \tau/4)) \end{aligned}\right\} \text{as } \operatorname{Im} \tau \longrightarrow \infty$$

This completes the proof of the proposition. (In fact, a similar reasoning shows that the analytic functions

$$\prod_{1 \le i \le 2\ell} \vartheta_{a_i b_i}(0, k_i\tau)\,, \quad a_i, b_i, k_i \in \mathbb{Q}, k_i > 0$$

are modular forms of weight $\ell$ and suitable level. We will prove this in a more general context below). The above proof also allows us to point out the following:

Remark 9.3. The modular forms $\vartheta_{00}^2(0,\tau)$, $\vartheta_{01}^2(0,\tau)$ and $\vartheta_{10}^2(0,\tau)$ look like

$$\zeta/(c\tau+d) + (\text{Error term})$$

when $\tau \longrightarrow -d/c$, where $\zeta = 0$ or $\zeta^8 = 1$. Here $\tau$ should approach $-d/c$ in horocircles of decreasing radii touching the real axis at $-d/c$, and the error term goes to 0 exponentially with the radius of the horocircle: more precisely,

$$(\text{Error term}) = O(\exp[-\text{constant}.\operatorname{Im}\tau/|c\tau+d|^2]).$$

(This is seen by estimating $f^{\gamma}(\tau)$ as $\tau \longrightarrow -d/c$ in terms of $f(\tau)$ as $\tau \longrightarrow \infty$).

Another simple fact which follows from the discussion above and which we need later is the following:

Remark 9.4. Let $f \in \mathrm{Mod}_k^{(N)}$. Then

$$f(\tau) = O((\mathrm{Im}\ \tau)^{-k}) \qquad \text{as } \mathrm{Im}\ \tau \longrightarrow 0$$

To see this: let us recall the classical fundamental domain $F$ for the action of $SL(2,\mathbb{Z})$ on $H$, namely,

$$F = \{\tau \in H / |\tau| \geq 1 \text{ and } |\mathrm{Re}\ \tau| \leq \tfrac{1}{2}\}$$

(cf. Fig. 3, § 10, below). So we have $H = \bigcup_{\gamma} \gamma F$, $\gamma \in SL(2,\mathbb{Z})$. Let

$$F' = \{\tau \in H / \mathrm{Im}\ \tau \geq 3^{\frac{1}{2}}/2\}.$$

Since $F \subset F'$, we have $H = \bigcup_{\gamma} \gamma F'$, $\gamma \in SL(2,\mathbb{Z})$. Take any $\tau \in H$. Then $\exists\ \gamma \in SL(2,\mathbb{Z})$ such that $\mathrm{Im}\,(\gamma\tau) \geq 3^{\frac{1}{2}}/2$. Moreover if $\gamma = \begin{pmatrix} a & b \\ c & d \end{pmatrix}$, either $\mathrm{Im}\ \tau \geq 3^{\frac{1}{2}}/2$ or $c \neq 0$. Now we make the following:

Claim: $\exists$ a constant $C_o > 0$ such that $|f(\tau)| \leq C_o |c\tau + d|^{-k}$ whenever $\gamma = \begin{pmatrix} a & b \\ c & d \end{pmatrix} \in SL(2,\mathbb{Z})$ is such that $\mathrm{Im}\,(\gamma\tau) \geq 3^{\frac{1}{2}}/2$.

Observe that this claim proves the remark because $c \neq 0$ implies

$$|c\tau + d| \geq |c.\ \mathrm{Im}\ \tau| = |c|\ .\ \mathrm{Im}\ \tau \geq \mathrm{Im}\ \tau,$$

i.e., $|f(\tau)| \leq C_o\,(\mathrm{Im}\ \tau)^{-k}$ for $\mathrm{Im}\ \tau < 3^{\frac{1}{2}}/2$, as asserted. To prove the claim: since $f \in \mathrm{Mod}_k^{(N)}$, we have:

$$f^{\gamma}(\tau) = (c\tau + d)^{-k} f(\gamma\tau),\ f^{\gamma} \in \mathrm{Mod}_k^{(N)} \text{ and } f^{\gamma\gamma'} = f^{\gamma},\ \forall\ \gamma' \in \Gamma_N\ .$$

In particular, there are only finitely many modular forms of the type $f^{\gamma}$, $\gamma \in SL(2,\mathbb{Z})$. Hence (by Def. 9.1, b(i)) $\exists$ a constant $C_o > 0$ such that $\forall \gamma \in SL(2,\mathbb{Z})$, we have:

(*) $$\left|f^{\gamma}(\tau)\right| \le C_o \quad \text{if } \operatorname{Im}\tau \ge 3^{\frac{1}{2}}/2$$

On the other hand, we have (by definition of $f^{\gamma}$) :

(**) $$f^{\gamma^{-1}}(\gamma\tau) = e_{\gamma^{-1}}(\gamma\tau)^{-1} f(\tau) = e_{\gamma}(\tau)f(\tau) = (c\tau+d)^k f(\tau).$$

Clearly then (*) and (**) imply the claim.

## § 10. The geometry of modular forms

Just as a set of theta functions with characteristics enabled us to embed $\mathbb{C}/\Lambda_\tau$ in a projective space, so we can take more than one "theta-null" $\vartheta_{ab}(0,\tau)$ and use these to embed $H/\Gamma_N$ in a projective space. We will look only at the simplest case $N = 4$. Then we saw that $\vartheta_{00}^2(0,\tau), \vartheta_{01}^2(0,\tau)$ & $\vartheta_{10}^2(0,\tau)$ were modular forms of weight 1 and level 4. Recall also from § 1 that $\frac{1}{2}(1+\tau)$ is the only zero of $\vartheta_{00}(z,\tau)$, so $\vartheta_{00}(0,\tau)$, $\vartheta_{01}(0,\tau)$ & $\vartheta_{10}(0,\tau)$ are never zero. It follows that we have a holomorphic map

$$\Psi_2 : H/\Gamma_4 \longrightarrow \mathbb{P}^2$$

defined by

$$\Psi_2(\tau) = (\vartheta_{00}^2(0,\tau), \vartheta_{01}^2(0,\tau), \vartheta_{10}^2(0,\tau)).$$

As in § 4, the point is that in $\Psi_2(\gamma\tau)$, each function picks up the same factor $e_\gamma(\tau)$ and so $\Psi_2$ is well-defined. Moreover, $\Psi_2$ is equivariant for the finite group $SL(2,\mathbb{Z})/\Gamma_4$ which acts on $H/\Gamma_4$ because $\Gamma_4$ is normal in $SL(2,\mathbb{Z})$, and on $\mathbb{P}^2$ because the 3 functions $\vartheta_{ij}^2(0,\tau)$ are mapped into combinations of themselves by every $\gamma \in SL(2,\mathbb{Z})$. In fact, using the action tabulated in §9, we find: if $\Psi_2(\tau) = (x_0, x_1, x_2)$, then

$$\Psi_2(\tau+1) = (x_1, x_0, ix_2) \text{ and } \Psi_2(-\tfrac{1}{\tau}) = (x_0, x_2, x_1)$$

Moreover, by equation ($J_1$) in §5, the image of $\Psi_2$ lies on the conic

$$A : x_0^2 = x_1^2 + x_2^2$$

but missing the 6 points $(1,0,\pm 1)$, $(1,\pm 1,0)$, $(0,1,\pm i)$ where the conic meets the coordinate axes $x_i = 0$. The missing points are clearly accounted for by the cusps: in fact, if $\operatorname{Im}\tau \longrightarrow +\infty$, then by the Fourier expansions of $\vartheta_{ij}(0,\tau)$, it is clear that

$$\vartheta_{00}(0,\tau) \longrightarrow +1,\ \vartheta_{01}(0,\tau) \longrightarrow +1 \text{ and } \vartheta_{10}(0,\tau) \longrightarrow 0$$

hence $\Upsilon_2(\tau) \longrightarrow (1,1,0)$. Acting by $SL(2,\mathbb{Z})$, the other cusps will map onto the other missing points. The easiest way to "extend $\Upsilon_2$ to the cusps" is this:

(a) define explicitly by "scissors & glue" a compactification $\widetilde{H/\Gamma_4}$ of the orbit space $H/\Gamma_4$ ; $\widetilde{H/\Gamma_4}$ is an abstract Riemann surface, and then

(b) verify the $\Upsilon_2$ extends to a holomorphic map on all of $\widetilde{H/\Gamma_4}$ .

For (a), it is useful to recall the classical fundamental domain for the action of $SL(2,\mathbb{Z})$ on $H$: viz., the set $F \subset H$ defined by

$$F = \{\tau \in H / |\tau| \geq 1 \text{ and } |\operatorname{Re}\tau| \leq \tfrac{1}{2}\}$$

(cf. diagram below).

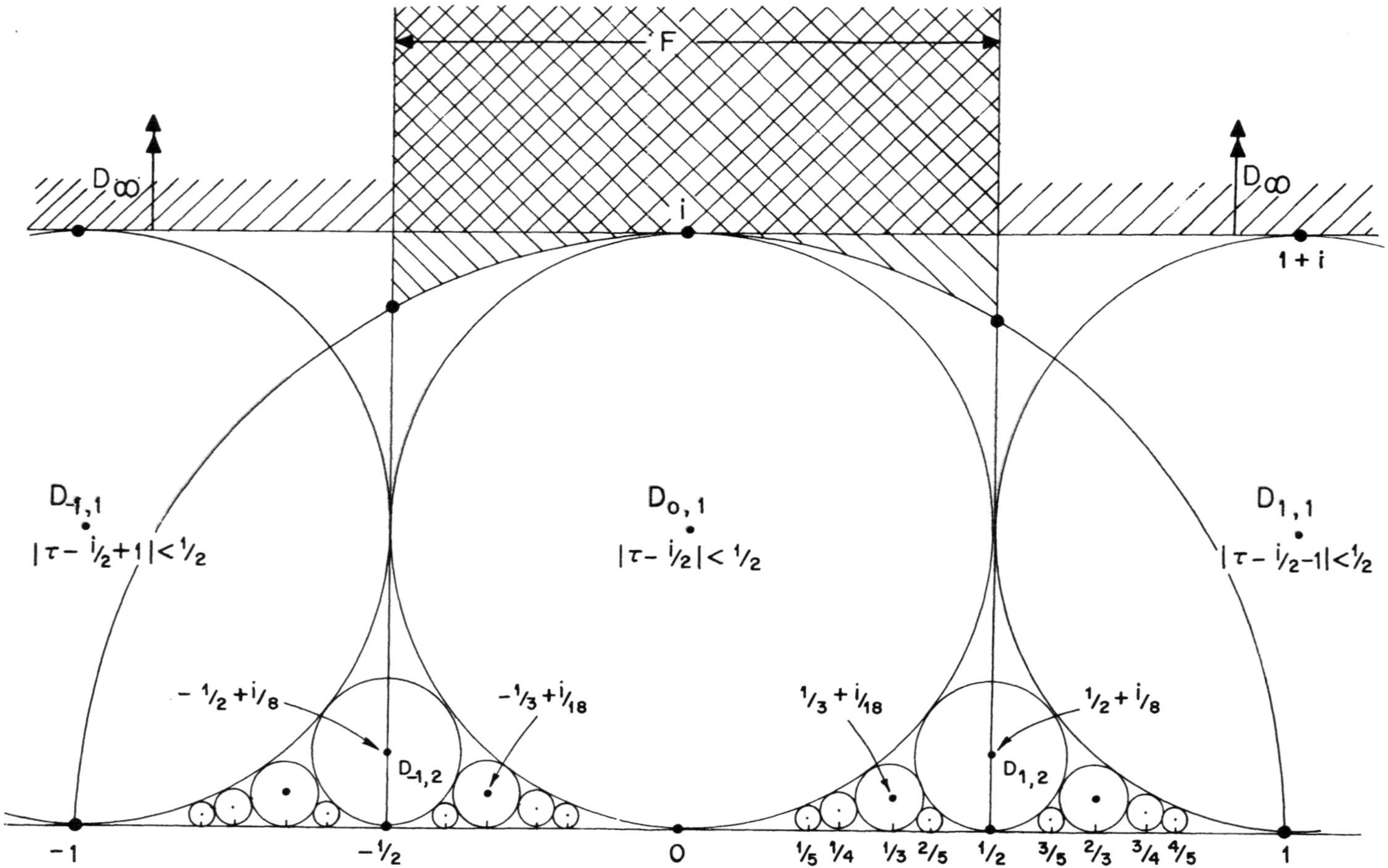


Fig. 3 Fundamental domain for SL(2,Z) and neighbourhoods of cusps.

Let

$$D_\infty = \{\tau \in H / \operatorname{Im} \tau > 1\}.$$

Since

$$D_\infty \subset \bigcup_{b \in \mathbb{Z}} (F + b) = \bigcup_{\gamma = \begin{pmatrix}1 & b\\ 0 & 1\end{pmatrix}} \gamma F ,$$

it follows that :

$\forall \tau_1, \tau_2 \in D_\infty$, if $\tau_1 \; \gamma\tau_2$ for some $\gamma \in SL(2, \mathbb{Z})$, then

$\tau_1 = \tau_2 + b, \gamma = \begin{pmatrix}1 & b\\ 0 & 1\end{pmatrix}$ and $\gamma(\infty) = \infty$ .

Thus mapping $D_\infty$ to $H/\Gamma_4$ identifies only the pairs of points $\tau$ and $\tau + 4k$, $k \in \mathbb{Z}$. Equivalently, if $w = \exp(\frac{1}{2}\pi i \tau)$, then

$$\begin{aligned} H/\Gamma_4 \supset \text{the punctured disc} &= \{w / 0 < |w| < \exp(-\tfrac{1}{2}\pi)\} \\ &= D_\infty / \{\begin{pmatrix}1 & b\\ 0 & 1\end{pmatrix} / b \equiv 0 \pmod 4\} . \end{aligned}$$

What we want to do is to glue together $H/\Gamma_4$ and the full disc $|w| \leq \exp(-\frac{1}{2}\pi)$, identifying these two on the punctured disc. We can do the same thing at the other cusps. For all $p/q \in \mathbb{Q}$, let $D_{p/q}$ be the horocircle:

$$D_{p/q} = \{\tau \in H / |\tau - p/q - i/2q^2| < 1/2q^2\}.$$

Then it is [redacted] easy to check that

$$\gamma(D_\infty) = D_{p/q} \text{ whenever } \gamma(\infty) = p/q,\ \gamma \in SL(2, \mathbb{Z}).$$

Thus, if $\tau_1, \tau_2 \in D_{p/q}$ and $\tau_1 = \delta\tau_2$ for some $\delta \in SL(2, \mathbb{Z})$, then $\delta$ is conjugate to $\begin{pmatrix}1 & b\\ 0 & 1\end{pmatrix}$, or what is the same,

$$\delta = \begin{pmatrix} 1 - bpq & bp^2 \\ -bq^2 & 1 + bpq \end{pmatrix} \text{ for some } b \in \mathbb{Z} .$$

Let $w_{p,q} : H \longrightarrow \mathbb{C}$ be the function defined by

$$w_{p,q}(\tau) = \exp(-\pi i/2q\,(q\tau - p)).$$

It is easy to see that : $\forall \tau_1, \tau_2 \in D_{p/q}$,

$$w_{p,q}(\tau_1) = w_{p,q}(\tau_2) \Longleftrightarrow \tau_1 = \delta\tau_2 \text{ for some } \delta \text{ as above.}$$

Thus, as before, the image of $D_{p/q}$ in $H/\Gamma_4$ identifies only the pairs $\tau$ and $\delta\tau$ for $\delta$ as above with $b \equiv 0 \pmod 4$. In other words, we have:

$$H/\Gamma_4 \supset \text{the punctured disc} = \{w_{p,q} / 0 < |w_{p,q}| < \exp(-\tfrac{1}{2}\pi)\}$$

$$= \text{Horocircle } D_{p/q} / \{\begin{pmatrix} 1-bpq & bp^2 \\ -bq^2 & 1+bpq \end{pmatrix} / b \equiv 0 \pmod 4\}.$$

Again, we glue the full disc $|w_{p,q}| < \exp(-\frac{1}{2}\pi)$ to $H/\Gamma_4$ along the punctured disc. Clearly, if $\gamma(p/q) = p'/q'$, $\gamma \in \Gamma_4$, then the above operation at $p/q$ or $p'/q'$ has the same effect on $H/\Gamma_4$. So we need only do it once for each orbit of $\Gamma_4$ acting on $\mathbb{Q} \cup \{\infty\}$. It is not hard to check that $\Gamma_4$ has 6 orbits on $\mathbb{Q} \cup \{\infty\}$, namely:

(i) $\Gamma_4(\infty) = \{\infty\} \cup \{p/4q \mid p \text{ odd}\}$

(ii) $\Gamma_4(0) = \{4p/q \mid q \text{ odd}\}$

(iii) $\Gamma_4(\frac{1}{2}) = \{\frac{1}{2}p/q \mid p, q \text{ odd}\}$

(iv) $\Gamma_4(1) = \{p/q \mid p, q \text{ odd and } p \equiv q \pmod 4\}$

(v) $\Gamma_4(2) = \{2p/q \mid p, q \text{ odd}\}$

(vi) $\Gamma_4(3) = \{p/q \mid p, q \text{ odd and } p \equiv -q \pmod 4\} = \Gamma_4(-1)$

So we define $\widetilde{H/\Gamma_4}$ to be $H/\Gamma_4$ with 6 "cusps" adjoined by the above procedure, one for each of the above orbits. It is á priori not at all obvious that $\widetilde{H/\Gamma_4}$

is a compact Hausdorff space! Perhaps, the easiest way to see this is to describe it alternatively by a fundamental domain: for $1 \leq i \leq 6$, let

$\gamma_i \in SL(2,\mathbb{Z})$ carry $\infty$ to the 6 cusps $\infty$, 0, $\frac{1}{2}$, 1, 2 & 3. For instance, we take $\gamma_i$ to be :

$$\begin{pmatrix}1 & 0\\ 0 & 1\end{pmatrix}, \begin{pmatrix}0 & -1\\ 1 & -2\end{pmatrix}, \begin{pmatrix}1 & -3\\ 2 & -5\end{pmatrix}, \begin{pmatrix}1 & -2\\ 1 & -1\end{pmatrix}, \begin{pmatrix}2 & -3\\ 1 & -1\end{pmatrix} \text{ and } \begin{pmatrix}3 & -4\\ 1 & -1\end{pmatrix}.$$

Let $\delta_j = \begin{pmatrix}1 & j\\ 0 & 1\end{pmatrix}$, $1 \leq j \leq 4$, be coset representatives in $G_\infty$ for the subgroup $G_\infty \cap \Gamma_4$ where $G_\infty$ is the stabiliser $\{\begin{pmatrix}1 & b\\ 0 & 1\end{pmatrix}\}$ of $\infty$ in $SL(2,\mathbb{Z})$. Then the 24 elements $\gamma_i \delta_j$ are coset representatives for $SL(2,\mathbb{Z})/\Gamma_4$. Therefore, $H/\Gamma_4$ is just the non-Euclidean polygon

$$\bigcup_{i,j} \gamma_i \delta_j F, \quad \begin{cases} 1 \leq i \leq 6 \\ 1 \leq j \leq 4 \end{cases}$$

with its edges identified in pairs (cf. Fig. 4 (*)). Clearly, the closure of this polygon meets the boundary $\mathbb{R} \cup \{\infty\}$ at the 6 cusps $\infty, 0, \frac{1}{2}, 1, 2$ & 3, and we have added these limit points to $H/\Gamma_4$ to obtain $\overline{H/\Gamma_4}$. Thus $\overline{H/\Gamma_4}$ is a compact Hausdorff space.

It is now easy to extend $\Psi_2$ to $\tilde{\Psi}_2 : \overline{H/\Gamma_4} \longrightarrow \mathbb{P}^2$: in fact, at the cusp at $\infty$, $w = \exp(\frac{1}{2}\pi i \tau)$ is the local coordinate on $\overline{H/\Gamma_4}$, and we have:

$$\vartheta_{00}(0,\tau) = \sum_{n\in\mathbb{Z}} \exp(\pi i n^2 \tau) = 1 + 2 \sum_{n\in\mathbb{N}} w^{2n^2}$$

$$\vartheta_{01}(0,\tau) = \sum_{n\in\mathbb{Z}} \exp(\pi i n^2 \tau + \pi i n) = 1 + 2 \sum_{n\in\mathbb{N}} (-1)^n w^{2n^2}$$

$$\vartheta_{10}(0,\tau) = \sum_{n\in\mathbb{Z}} \exp(\pi i (n+\tfrac{1}{2})^2 \tau) = 2 w^{\frac{1}{2}} \sum_{n\in\mathbb{Z}^+} w^{2(n^2+n)}$$

(*) In thinking about these diagrams in the non-Euclidean plane, it is good to bear in mind a comment of Thurston: these diagrams make it look like the space gets very crowded and hot near the boundary; in reality, however, the space is increasingly empty and quite cold near the boundary.

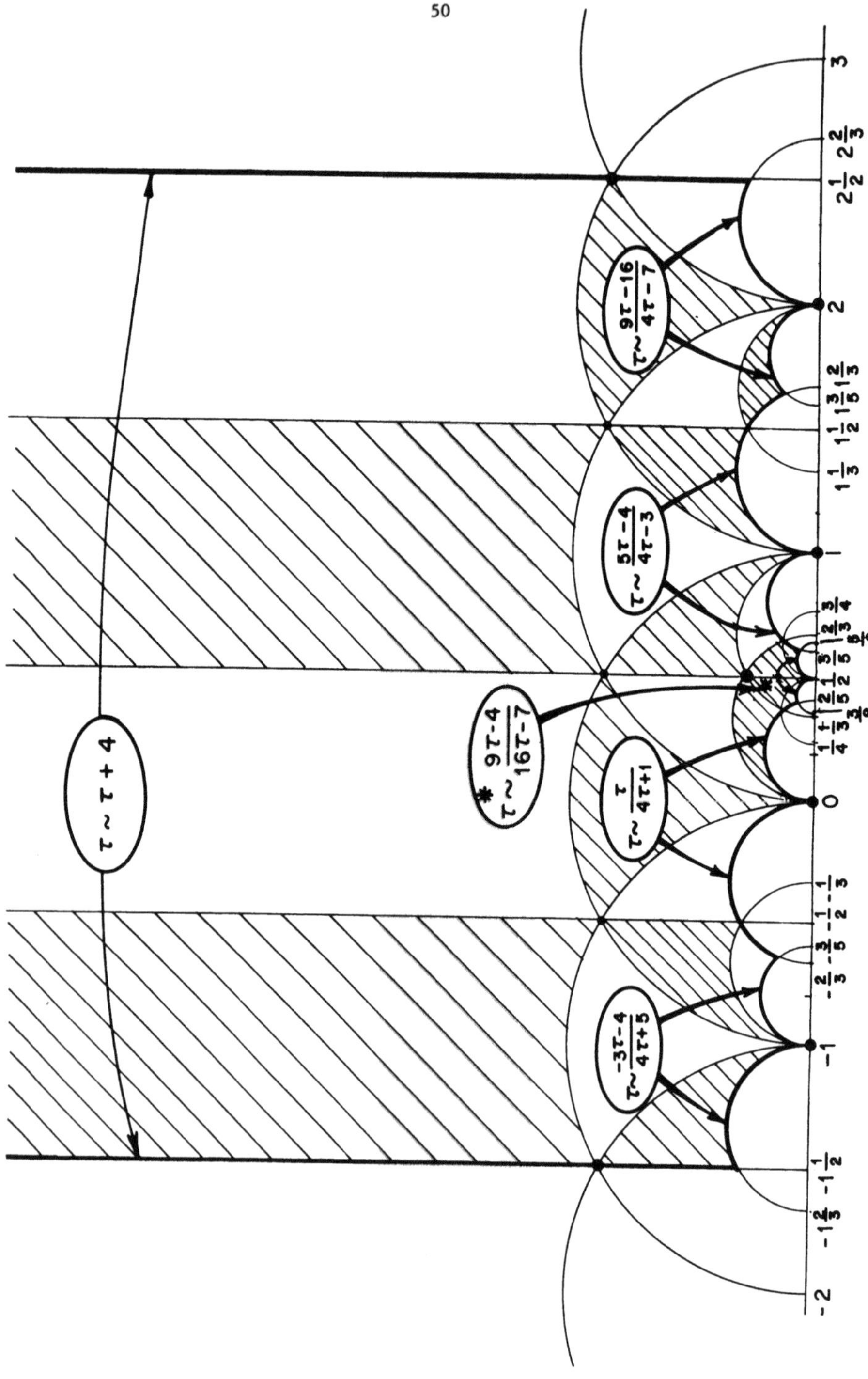


Fig. 4 Fundamental Domain for $\Gamma_4$

[ hence $\vartheta_{10}^2(0,\tau) = 4\,w\big(\sum_{n\in\mathbb{Z}^+} w^{2(n^2+n)}\big)^2$ ].

Thus $\tilde{\Psi}_2$ is holomorphic in $w$, carrying the cusp $w = 0$ to $(1,1,0) \in \mathbb{P}^2$. But then it follows, by $SL(2,\mathbb{Z})$ - equivariance that $\tilde{\Psi}_2$ is holomorphic at the other cusps too. Finally, we have the simple:

Theorem 10.1. The naturally extended holomorphic map

$$\tilde{\Psi}_2 : H/\tilde{\Gamma}_4 \longrightarrow [\text{conic } A: x_0^2 = x_1^2 + x_2^2 ]$$

is an isomorphism.

Proof. In fact, both $H/\tilde{\Gamma}_4$ and $A$ are compact Riemann surfaces, and $\tilde{\Psi}_2$ is a non-constant holomorphic map. Therefore, $\tilde{\Psi}_2$ makes $H/\tilde{\Gamma}_4$ a (possibly) ramified covering of $A$. To see that $\tilde{\Psi}_2$ is an isomorphism, we need only check that its degree is 1. But if its degree is d, then over each point of $A$, there are d points (counted with multiplicities where $\tilde{\Psi}_2$ is ramified). Now consider the 6 points $(1, \pm 1, 0)$, $(1, 0, \pm 1)$ and $(0, 1, \pm i)$. Only cusps can be mapped to these and there are 6 cusps. Thus only the cusp $\tau = i\infty$ is mapped to $(1,1,0)$. But by the formulae above, we have:

$$\frac{d}{dw}\big(\vartheta_{10}^2(0,\tau)/\vartheta_{00}^2(0,\tau)\big)\Big|_{w=0} \neq 0$$

which means that $\tilde{\Psi}_2$ is unramified at $i\infty$. Hence degree of $\tilde{\Psi}_2$ is 1, i.e., $\tilde{\Psi}_2$ is an isomorphism. (In fact, it can be checked with the formulae we have at hand that the cusps $\infty, 0, \frac{1}{2}, 1, 2$ & $3$ are respectively mapped to the points $(1,1,0)$, $(1,0,1)$, $(1,-1,0)$, $(0,1,i)$, $(1,0,-1)$ and $(0,1,-i)$.

An important consequence of this theorem is:

Corollary 10.2. The ring $\mathrm{Mod}^{(4)}$ of modular forms of level 4 is naturally isomorphic to

$$\mathbb{C}[\vartheta_{00}^2(0,\tau), \vartheta_{01}^2(0,\tau), \vartheta_{10}^2(0,\tau)]/(\vartheta_{00}^4 - \vartheta_{01}^4 - \vartheta_{10}^4)$$

i.e., it is generated by $\vartheta_{ij}^2(0,\tau)$ and subject to only the relation $(J_1)$.

Proof. Let $f \in \mathrm{Mod}_k^{(4)}$. Then $f/\vartheta_{00}^{2k}(0,\tau)$ is a meromorphic function on $H/\tilde{\Gamma}_4$ with poles only where $\vartheta_{00}(0,\tau) = 0$, i.e., only at the 2 cusps 1 and 3, and there poles of order at most k (recall that just as $\vartheta_{10}^2(0,\tau)$ has a simple zero at $\tau = i\infty$, so also $\vartheta_{00}^2(0,\tau)$ has a simple zero at 1 and 3). Therefore, it corresponds to a meromorphic function $g$ on the conic $A$ with at most $k$-fold poles at the points $(0, 1, \pm i)$. But $A$ is biholomorphically isomorphic to the projective line $\mathbb{P}^1$ via the map:

$$x_0 \longmapsto t_0^2 + t_1^2,\ x_1 \longmapsto 2t_0t_1 \text{ and } x_2 \longmapsto t_0^2 - t_1^2$$

where $(t_0, t_1)$ are homogeneous coordinates on $\mathbb{P}^1$. Here $t_0 = 1$ and $t_1 = \pm i$ correspond to the points $(x_0, x_1, x_2) = (0, 1, \mp i)$. So $g$ corresponds to a meromorphic function $h$ on $\mathbb{P}^1$ with $k$-fold poles at $t_0 = 1$, $t_1 = \pm i$. Hence $h$ is a rational function of $t_1/t_0$ and by partial fraction decomposition of rational functions, one checks easily that it can be written as:

$$h = Q(t_0, t_1)/(t_0^2 + t_1^2)^k$$

for some homogeneous polynomial Q of degree 2k. Thus

$$g = P(x_0, x_1, x_2)/x_0^k$$

for some $P$ homogeneous of degree $k$. Thus

$$f/\vartheta_{00}^{2k}(0,\tau) = P(\vartheta_{00}^2, \vartheta_{01}^2, \vartheta_{10}^2)/\vartheta_{00}^{2k}(0,\tau).$$

Finally, there can be no further relations between $\vartheta_{00}^2, \vartheta_{01}^2, \vartheta_{10}^2$ because the only polynomials that vanish on the conic $x_0^2 = x_1^2 + x_2^2$ are multiples of $x_0^2 - x_1^2 - x_2^2$, as required.

## § 11. $\vartheta$ as an automorphic form in 2 variables.

So far we have concentrated on the behaviour of $\vartheta(z,\tau)$ as a function of $z$ for fixed $\tau$, and as a function of $\tau$ for $z = 0$. Let us now put all this together and consider $\vartheta$ as a function of both variables. First of all, it is easy to see that the functional equations on $\vartheta$, plus its limiting behaviour as $\operatorname{Im}\tau \longrightarrow \infty$ characterise $\vartheta$ completely. More precisely:

Proposition 11.1. $\vartheta(z,\tau)$ is the unique holomorphic function $f(z,\tau)$ on $\mathbb{C} \times H$ such that

a) $f(z+1, \tau) = f(z,\tau)$

b) $f(z+\tau, \tau) = \exp(-\pi i\tau - 2\pi iz) \,.\, f(z,\tau)$

c) $f(z+\frac{1}{2}, \tau+1) = f(z,\tau)$

d) $f(z/\tau, -1/\tau) = (-i\tau)^{\frac{1}{2}} \exp(\pi iz^2/\tau) \,.\, f(z,\tau)$

and for all $z \in \mathbb{C}$,

e) $\lim_{\operatorname{Im}\tau \longrightarrow +\infty} f(z,\tau) = 1.$

Proof. We have used all these properties of $\vartheta(z,\tau)$ repeatedly, except perhaps (c) which follows from the identity:

$$\vartheta(z+\tfrac{1}{2}, \tau+1) = \sum_{n\in\mathbb{Z}} \exp[\pi i n^2(\tau+1) + 2\pi i n(z+\tfrac{1}{2})]$$

$$= \sum_{n\in\mathbb{Z}} (-1)^{n^2+n} \exp(\pi i n^2\tau + 2\pi i n z)$$

$$= \vartheta(z, \tau).$$

Conversely, to see that these properities characterise $\vartheta(z, \tau)$, take any such f: by the results of § 1, (a) and (b) imply that

$$f(z, \tau) = g(\tau)\,\vartheta(z, \tau)$$

for some holomorphic function $g(\tau)$ on H. Now by the results of § 7, (c) and (d) imply that

$$g(\tau+1) = g(\tau) \;\&\; g(-\tau^{-1}) = g(\tau).$$

Thus $g(\tau)$ is a holomorphic function on $H/SL(2,\mathbb{Z})$. On the other hand, (e) implies that $g(\tau) \longrightarrow 1$ as $\operatorname{Im}\tau \longrightarrow +\infty$. This means that $g(\tau)$ is bounded outside a horizontal strip and hence by $SL(2,\mathbb{Z})$-invariance, it is bounded everywhere. Thus $|g(\tau)-1|$, if not identically zero, takes a positive maximum at some point of F which cannot happen. So $g(\tau) \equiv 1$, as required.

The 4 theta functions $\vartheta_{ij}(z, \tau)$ moreover satisfy together a system of functional equations that we have given in Table 0, § 5 and Table V, § 9. To understand the geometric implications of these, we consider the holomorphic map:

$$\Phi : \mathbb{C}\times H \longrightarrow \mathbb{P}^3, (z,\tau) \longmapsto (\vartheta_{00}(2z,\tau), \vartheta_{01}(2z,\tau), \vartheta_{10}(2z,\tau), \vartheta_{11}(2z,\tau)).$$

The semi-direct product

$$(\tfrac{1}{4}\mathbb{Z})^2 \ltimes SL(2, \mathbb{Z})$$

acts on $\mathbb{C}\times H$ by

$$(m,n;\begin{pmatrix} a & b \\ c & d \end{pmatrix}) : (z, \tau) \longmapsto \left(\frac{z+m\tau+n}{c\tau+d}, \frac{a\tau+b}{c\tau+d}\right),$$

and we have seen (by the tables referred to above) that the 4 generators $(\frac{1}{4}, 0; I)$, $(0, \frac{1}{4}; I)$, $(0,0; \begin{pmatrix} 1 & 1 \\ 0 & 1 \end{pmatrix})$ and $(0,0; \begin{pmatrix} 0 & -1 \\ 1 & 0 \end{pmatrix})$ of this group transform the 4 functions $\vartheta_{ij}(2z, \tau)$ into themselves. In other words, the map $\Phi$ is equivariant when the same group acts on $\mathbb{P}^3$ via:

$$(\tfrac{1}{4}, 0; I) : (x_0, x_1, x_2, x_3) \longmapsto (x_2, -ix_3, x_0, -ix_1)$$
$$(0, \tfrac{1}{4}; I) : (\quad '' \quad) \longmapsto (x_1, x_0, x_3, -x_2)$$
$$(0, 0; \begin{pmatrix} 1 & 1 \\ 0 & 1 \end{pmatrix}) : (\quad '' \quad) \longmapsto (x_1, x_0, \lambda x_3, \lambda x_2) \text{ where } \lambda = \exp(\pi i/4)$$
$$(0, 0; \begin{pmatrix} 0 & -1 \\ 1 & 0 \end{pmatrix}) : (\quad '' \quad) \longmapsto (x_0, x_2, x_1, -ix_3).$$

Now we have the following:

**Proposition 11.2.** Let $\Gamma^* \subset (\frac{1}{4}\mathbb{Z})^2 \ltimes SL(2,\mathbb{Z})$ be defined by

$$\Gamma^* = \{(m,n;\begin{pmatrix} a & b \\ c & d \end{pmatrix}) / \begin{pmatrix} a & b \\ c & d \end{pmatrix} \in \Gamma_4, m \equiv \tfrac{c}{8} \pmod 1 \ \&\ n \equiv \tfrac{b}{8} \pmod 1\}.$$

Then $\Gamma^*$ is a normal subgroup of $(\frac{1}{4}\mathbb{Z})^2 \ltimes SL(2,\mathbb{Z})$, it acts trivially on $\mathbb{P}^3$ and :

$$\Phi(z,\tau) = \Phi(z',\tau') \Longleftrightarrow (z',\tau') = \gamma(z,\tau) \text{ for some } \gamma \in \Gamma^*.$$

Thus $\Phi$ collapses the action of $\Gamma^*$ on $\mathbb{C} \times H$ and carries $(\mathbb{C} \times H)/\Gamma^*$ into the quartic surface $F$ in $\mathbb{P}^3$ defined by

$$F : x_0^4 + x_3^4 = x_1^4 + x_2^4 .$$

*Proof.* We give the proof in 6 steps:

(1). That $\Gamma^*$ is a normal subgroup of $(\frac{1}{4}\mathbb{Z})^2 \ltimes SL(2, \mathbb{Z})$ is a straightforward verification using the fact that $\begin{pmatrix} a & b \\ c & d \end{pmatrix} \longmapsto \frac{b}{4}, \frac{c}{4}$ are homomorphisms from $\Gamma_4 \longrightarrow \mathbb{Z}/2\mathbb{Z}$.

(2) Note that $\Gamma_4$ is the least normal subgroup of $SL(2, \mathbb{Z})$ containing $\begin{pmatrix} 1 & 4 \\ 0 & 1 \end{pmatrix}^{(*)}$ : in fact, if N is the least normal subgroup in question, then the fact that $N = \Gamma_4$ can be seen in several ways, viz.,

(i) Topological way: look at the fundamental domain for $H/\Gamma_4$ (cf. Fig.4): let $\gamma_i \in \Gamma_4$, $1 \leq i \leq 6$, be the transformations identifying in pairs of the edges of the diagram, namely,

$$\gamma_i = \begin{pmatrix} 1 & -4 \\ 0 & 1 \end{pmatrix}, \begin{pmatrix} -3 & -4 \\ 4 & 5 \end{pmatrix}, \begin{pmatrix} 1 & 0 \\ 4 & 1 \end{pmatrix}, \begin{pmatrix} 9 & -4 \\ 16 & -7 \end{pmatrix}, \begin{pmatrix} 5 & -4 \\ 4 & -3 \end{pmatrix} \text{ and } \begin{pmatrix} 9 & -16 \\ 4 & -7 \end{pmatrix}.$$

It follows that $\Gamma_4$ is generated as a group by these $\gamma_i$'s. On the other hand, it is easy to see that the $\gamma_i$, $2 \leq i \leq 6$, are conjugates of $\gamma_1 = \begin{pmatrix} 1 & -4 \\ 0 & 1 \end{pmatrix}$ and hence $N = \Gamma_4$.

(ii) Abstract way: recall that $SL(2, \mathbb{Z})$ is generated by $a = \begin{pmatrix} 1 & 1 \\ 0 & 1 \end{pmatrix}$ and $b = \begin{pmatrix} 0 & -1 \\ 1 & 0 \end{pmatrix}$ and hence their residues mod N generate $SL(2, \mathbb{Z})/N$, and modulo N we find that

$$a^4 = b^4 = 1 \text{ and } b^2 = (ab)^3 = (ba)^3$$

Clearly then $b^2$ is in the centre and, mod $b^2$ we have

$$a^4 = b^2 = (ab)^3 = 1.$$

But this is a well-known presentation of the Octahedral group of order 24 (cf. e.g., Coxeter-Moser, Appendix Table I). Thus

---

(*) But $\Gamma_n$ is not in general the least normal subgroup N containing $\begin{pmatrix} 1 & n \\ 0 & 1 \end{pmatrix}$!
In fact, for $n \geq 6$, N is not even of finite index!!

$$\#SL(2,\mathbb{Z})/N = 48 = \# SL(2,\mathbb{Z})/\Gamma_4 .$$

But $N \subseteq \Gamma_4$ and hence $N = \Gamma_4$ , as required.

Combining the facts (1) and (2), we get:

(3) . $\Gamma^*$ is the least normal subgroup of $(\frac{1}{4}\mathbb{Z})^2 \ltimes SL(2,\mathbb{Z})$ containing $\mathbb{Z}^2$ and $(0,\frac{1}{2},\begin{pmatrix}1 & 4\\ 0 & 1\end{pmatrix})$.

(4) $\Phi$ collapses the action of $\Gamma^*$: this is immediate since the same is true for the action of $\mathbb{Z}^2$ and $(0,\frac{1}{2};\begin{pmatrix}1 & 4\\ 0 & 1\end{pmatrix})$, as seen from the Tables O and V. Hence $\Phi$ factors through $(\mathbb{C}\times H)/\Gamma^*$.

(5). Suppose $\Phi(z,\tau) = \Phi(z',\tau') = P$, say. Recall from § 5 that for $\tau$ fixed, we have

$$\Phi(\mathbb{C}\times\{\tau\}) = \varphi_2(E_\tau) = C_\tau \subset \mathbb{P}^3$$

where $C_\tau$ is the elliptic curve defined by the equations:

$$(*)\qquad \begin{cases} a_o x_o^2 = a_1 x_1^2 + a_2 x_2^2 \\ a_o x_3^2 = a_2 x_1^2 - a_1 x_2^2 \end{cases}$$

$$(a_o, a_1, a_2) = (\vartheta_{oo}^2(0,\tau), \vartheta_{01}^2(0,\tau), \vartheta_{10}^2(0,\tau)).$$

Also, $C_\tau$ satisfies the further equations:

$$(*')\qquad \begin{cases} a_o x_1^2 = a_1 x_o^2 + a_2 x_3^2 \\ a_o x_2^2 = a_2 x_o^2 - a_1 x_3^2 \end{cases}$$

obtained by combining the 2 above equations. (For certain limiting values of the $a_i$'s, like $a_o = 0$, $a_1 = 1$, $a_2 = i$; the first two equations become dependent, but we can always find two independent ones in this full set of 4 equations).

By assumption $P \in C_\tau \cap C_{\tau'}$. Now look at the

Lemma 11.3. For all $\tau, \tau' \in H$, the curves $C_\tau$ and $C_{\tau'}$ are either identical or disjoint; in particular, we have (by § 10):

$$C_\tau \cap C_{\tau'} \neq \emptyset \Longleftrightarrow \Psi_2(\tau) = \Psi_2(\tau') \text{ in } \mathbb{P}^2$$
$$\Longleftrightarrow \tau' = \gamma(\tau) \text{ for some } \gamma \in \Gamma_4 .$$

Indeed, any point $P = (x_o, x_1, x_2, x_3)$ on the curve $C_\tau$ determines the curve completely because we can solve (*) for the $a_i$'s (upto scalars) in terms of the $x_i$'s obtaining:

$$(**) \qquad \begin{cases} a_o = \lambda(x_1^4 + x_2^4) \\ a_1 = \lambda(x_o^2 x_1^2 - x_2^2 x_3^2) \\ a_2 = \lambda(x_o^2 x_2^2 + x_1^2 x_3^2) \end{cases}$$

This gives the $a_i$'s in terms of the $x_i$'s unless all the expressions on the right are 0, e.g., when $x_1 = x_2 = 0$. However, if we solve the 1st equations in (*) and (*'), we get:

$$(**') \qquad \begin{cases} a_o = \mu(x_o^2 x_2^2 - x_1^2 x_3^2) \\ a_1 = \mu(x_1^2 x_2^2 - x_o^2 x_3^2) \\ a_2 = \mu(x_o^4 - x_1^4) \end{cases}$$

and there are no non-zero $x_i$'s for which all expressions on the right of (**) and (**') are zero, hence the lemma follows.

Coming back to the situation of step 5, we therefore get that $\tau' = \bar{\gamma}(\tau)$ for some $\bar{\gamma} \in \Gamma_4$. Now lift $\bar{\gamma}$ to $\gamma \in \Gamma^*$ and let $\gamma(z, \tau) = (z'', \tau')$. Then

$$\Phi(z', \tau') = \Phi(z'', \tau'), \text{ i.e., } \varphi_2(z') = \varphi_2(z'') \in C_{\tau'}$$

But $\varphi_2$ embeds the torus $E_{\tau'}$ in $\mathbb{P}^3$ and so we must have $z'-z'' \in \Lambda_{\tau'}$. Thus $\exists\, \delta \in \Gamma^*$ such that $\delta(z'', \tau') = (z', \tau')$ and hence $\delta\gamma(z, \tau) = (z', \tau')$, as required.

(6). Image $\Phi \subseteq$ Surface F : this is immediate beacuse squaring and adding the equations (*), we get

$$a_o^2(x_o^4 + x_3^4) = (a_1^2 + a_2^2)(x_1^4 + x_2^4) .$$

But from §5, Jacobi's identify $(J_1)$ gives $a_o^2 = a_1^2 + a_2^2$ implying $x_o^4 + x_3^4 = x_1^4 + x_2^4$, as required. This completes the proof of the proposition.

As an immediate consequence of Prop. 11.2, we deduce the following:

Corollary 11.4. The surface F in $\mathbb{P}^3$ has a fibre structure over the conic A in $\mathbb{P}^2$.

To see this; define a holomorphic map:

$$\pi : F \longrightarrow A, \ (x_o, x_1, x_2, x_3) \longmapsto (a_o, a_1, a_2)$$

by whichever of the 2 formulae that gives $(a_o, a_1, a_2) \neq (0, 0, 0)$, i.e.

$$\begin{aligned} a_o &= x_1^4 + x_2^4 \\ a_1 &= x_o^2x_1^2 - x_2^2x_3^2 \\ a_2 &= x_o^2x_2^2 + x_1^2x_3^2 \end{aligned} \qquad \text{or} \qquad \begin{aligned} a_o &= x_o^2x_2^2 - x_1^2x_3^2 \\ a_1 &= x_1^2x_2^2 - x_o^2x_3^2 \\ a_2 &= x_o^4 - x_1^4 \end{aligned}$$

(Using $x_o^4 + x_3^4 = x_1^4 + x_2^4$, we see that the 2 sets of formulae agree and that $a_o^2 = a_1^2 + a_2^2$). It is clear that the individual curves $C_\tau \subset F$ can be recovered as the inverse images under $\pi$ of the points $\Upsilon_2(\tau) \in A$. On the other hand, it is also clear that $\pi(F_o) = A_o$ where $F_o$ = Image $\Phi$ and $A_o$ = (A-6 cusps). In other words, $F_o$ is a fibre space formed out of the various curves $C_\tau$ lying over $A_o$. Further more, $F_o \approx (\mathbb{C} \times H)/\Gamma^*$.

We may summarise the discussion by the following commutative diagram:

$$\begin{array}{lccccccc}
 & \mathbb{C}\times H & \longrightarrow & (\mathbb{C}\times H)/\Gamma^* & \xrightarrow[\psi]{\sim} & F_o & \dashrightarrow & F\subset \mathbb{P}^3 \\
\text{fibres} = & \downarrow \mathbb{C} & & \downarrow E_\tau & & \downarrow C_\tau & & \downarrow \pi \\
 & H & \longrightarrow & H/\Gamma_4 & \xrightarrow[\gamma_2]{\sim} & A_o & \longrightarrow & A\subset \mathbb{P}^2
\end{array}$$

This suggests the interpretation of $H/\Gamma_4$ or $A_o$ as a moduli space which we will take up in the next section.

## § 12. Interpretation of $H/\Gamma_4$ as a moduli space

We are led to the interpretation of $H/\Gamma_4$ (or more generally $H/\Gamma_n$) as a moduli space when we ask: if $\tau, \tau' \in H$,

when are the complex tori $E_\tau$ and $E_{\tau'}$ biholomorphic?

Obviously, any biholomorphic map $f: E_\tau \longrightarrow E_{\tau'}$ lifts to their universal coverings $\mathbb{C}$, i;e., it is induced by a biholomorphic map $\tilde{f} : \mathbb{C} \longrightarrow \mathbb{C}$ such that (for $\lambda \in \Lambda_\tau$)

$$\tilde{f}(z+\lambda) = \tilde{f}(z) + f_*(\lambda)$$

where $f_* : \Lambda_\tau \longrightarrow \Lambda_{\tau'}$ is the isomorphism of the fundamental groups of $E_\tau$ and $E_{\tau'}$ induced by $f$. Then the derivative $\tilde{f}'$ is a doubly periodic entire function of $z$, hence it is a constant, i.e.,

$$\tilde{f}(z) = Lz + M$$

for some $L, M \in \mathbb{C}$ and therefore

$$L(z+\lambda) + M = Lz + M + f_*(\lambda)$$

i.e., $f_*(\lambda) = L\lambda$ and multiplication by $L$ is a bijection from $\Lambda_\tau$ onto

$\Lambda_{\tau'}$. In particular, $L \in \Lambda_{\tau'}$, or $L = c\tau' + d$ for some $c, d \in \mathbb{Z}$. Moreover, $L\tau \in \Lambda_{\tau'}$, i.e., $L\tau = a\tau' + b$ for some $a, b \in \mathbb{Z}$. Thus

$$\tau = \frac{a\tau' + b}{c\tau' + d}.$$

On the other hand, $L$ and $L\tau$ must generate $\Lambda_{\tau}$, which means $ad - bc = \pm 1$. But an easy calculation gives that

$$\operatorname{Im}\left(\frac{a\tau' + b}{c\tau' + d}\right) = \frac{(ad-bc)\operatorname{Im}\tau'}{|c\tau' + d|^2} = \operatorname{Im}\tau .$$

So since $\tau, \tau' \in H$, we must have $ad - bc = +1$. Thus we have:

$$E_\tau \approx E_{\tau'} \Longleftrightarrow \tau = \gamma\tau' \text{ for some } \gamma \in SL(2, \mathbb{Z}).$$

The converse is clear: if $\tau = (a\tau' + b)/(c\tau' + d)$, define $\tilde{f}(z) = (c\tau' + d)z$. Note that $\tilde{f}(\Lambda_\tau) = \Lambda_{\tau'}$ and hence $\tilde{f}$ induces an isomorphism $f: E_\tau \xrightarrow{\sim} E_{\tau'}$. Therefore, we have proved the well-known fact:

<u>Proposition 12.1.</u> Let $\tau, \tau' \in H$. Then $\tau = \gamma(\tau')$ for some $\gamma \in SL(2, \mathbb{Z})$ if and only if $\exists$ a biholomorphic map $f : E_\tau \longrightarrow E_{\tau'}$. Or, equivalently,

$$H/SL(2, \mathbb{Z}) \approx \left\{ \begin{array}{l} \text{Set of complex tori } E_\tau \text{ modulo} \\ \text{biholomorphic equivalence} \end{array} \right\}$$

<u>Now we ask: what then is the space</u> $H/\Gamma_n$ ?

Something stronger than "biholomorphic equivalence" is needed and it is done as follows: fix an $n$ and consider the 2 natural automorphisms of $\mathbb{C} \times H$, namely:

$$\alpha_n : (z, \tau) \longmapsto (z + \tfrac{1}{n}, \tau)$$

$$\beta_n : (z, \tau) \longmapsto (z + \tfrac{\tau}{n}, \tau)$$

Observe that for each $\tau$ fixed, $\alpha_n$ and $\beta_n$ induce automorphisms $\alpha_n^\tau, \beta_n^\tau$

of the torus $E_\tau$, i.e., $\alpha_n^\tau$ and $\beta_n^\tau$ are the translations on $E_\tau$ by the 2 generators of the group (of n-division points) $\frac{1}{n}\Lambda_\tau/\Lambda_\tau$. Now we have the following:

<u>Proposition 12.2.</u> Let $\tau, \tau' \in H$. Then $\tau = \gamma(\tau')$ for some $\gamma \in \Gamma_n$ if and only if $\exists$ a biholomorphic map $f: E_\tau \longrightarrow E_{\tau'}$ giving the commutative diagrams:

$$\begin{array}{ccc} E_\tau & \xrightarrow{f} & E_{\tau'} \\ \alpha_n^\tau \downarrow\downarrow \beta_n^\tau & & \alpha_n^{\tau'} \downarrow\downarrow \beta_n^{\tau'} \\ E_\tau & \xrightarrow{f} & E_{\tau'} \end{array} \qquad \text{i.e.,} \quad \begin{cases} \alpha_n^{\tau'} \circ f = f \circ \alpha_n^\tau \\ \beta_n^{\tau'} \circ f = f \circ \beta_n^\tau \end{cases}$$

Or, equivalently,

$$H/\Gamma_n \cong \left\{ \begin{array}{l} \text{Set of complex tori } E_\tau \text{ modulo isomorphisms preserving} \\ \text{the pair of automorphisms } \alpha_n^\tau \text{ and } \beta_n^\tau \text{ .} \end{array} \right\}$$

<u>Proof.</u> Let $f: E_\tau \longrightarrow E_{\tau'}$ be a biholomorphic map and $\gamma = \begin{pmatrix} a & b \\ c & d \end{pmatrix} \in SL(2,\mathbb{Z})$ be such that $\tau = \gamma(\tau')$. With L and M as above, we find that

$$\begin{aligned} f \circ \alpha_n^\tau = \alpha_n^{\tau'} \circ f &\Longleftrightarrow L(z+\tfrac{1}{n}) + M = (Lz+M) + \tfrac{1}{n} + \lambda, \ \lambda \in \Lambda_{\tau'} \\ &\Longleftrightarrow \frac{c\tau' + d - 1}{n} \in \Lambda_{\tau'} \\ &\Longleftrightarrow c, d-1 \equiv 0 \pmod n . \end{aligned}$$

Likewise,

$$\begin{aligned} f \circ \beta_n^\tau = \beta_n^{\tau'} \circ f &\Longleftrightarrow L(z+\tfrac{\tau}{n}) + M = (Lz+M) + \tfrac{\tau'}{n} + \mu \in \Lambda_{\tau'} \\ &\Longleftrightarrow \frac{(c\tau'+d)\,\tau - \tau'}{n} \in \Lambda_{\tau'} \\ &\Longleftrightarrow \frac{(a-1)\tau' + b}{n} \in \Lambda_{\tau'} \\ &\Longleftrightarrow a-1,\ b \equiv 0 \pmod n \end{aligned}$$

Thus both occur if and only if $\gamma \in \Gamma_n$, as required.

Let us look at the particular case $n = 4$: we constructed in the previous section a diagram:

$$\begin{array}{ccccc}
\mathbb{C}\times H & \longrightarrow & (\mathbb{C}\times H)/\Gamma_4^* & \xrightarrow[\Phi]{\sim} & F_o \subset F \subset \mathbb{P}^3 \\
\downarrow & & \downarrow & & \downarrow \qquad \downarrow \pi \\
H & \longrightarrow & H/\Gamma_4 & \xrightarrow[\Psi_2]{\sim} & A_o \subset A \subset \mathbb{P}^2
\end{array}$$

We can add to this diagram the auxiliary maps $\alpha_4$ & $\beta_4$ from $\mathbb{C}\times H$ to itself. If we define $\alpha_4'$ & $\beta_4'$ on $\mathbb{P}^3$ by

$$\alpha_4' : (x_o, x_1, x_2, x_3) \longmapsto (x_1, x_o, x_3, -x_2)$$

$$\beta_4' : (\qquad " \qquad) \longmapsto (x_2, -ix_3, x_o, -ix_1)$$

then $\Phi$ is equivariant, i.e., we have a commutative diagram:

$$\begin{array}{ccc}
(\mathbb{C}\times H)/\Gamma_4^* & \xrightarrow{\Phi} & F_o \\
\alpha_4 \downarrow\downarrow \beta_4 & & \alpha_4' \downarrow\downarrow \beta_4' \\
(\mathbb{C}\times H)/\Gamma_4^* & \xrightarrow{\Phi} & F_o
\end{array}$$

In other words, to each $a \in A_o$, we can associate the fibre $\pi^{-1}(a) \subset F_o$ plus 2 automorphisms res $\alpha_4'$ & res $\beta_4'$, and we have shown that distinct points $a \in A_o$ are associated to non-isomorphic triples $(\pi^{-1}a, \text{res } \alpha_4', \text{res } \beta_4')$. Thus $(F_o \longrightarrow A_o; \alpha_4', \beta_4')$ is a kind of universal family of complex tori, with 2 automorphisms of order 4 which sets up the set-theoretic bijection:

$$A_o \approx \left\{ \begin{array}{l} \text{Set of complex tori } E_\tau \text{ plus automorphisms} \\ \alpha_4^\tau, \beta_4^\tau \text{ modulo isomorphisms} \end{array} \right\} .$$

More details on this moduli interpretation can be found in Deligne-Rapopart, Les Schémas de modules de courbes elliptiques, in Springer Lecture Notes No. 349. Similar constructions can be carried through for all n, but the formulae are much more complicated.

## § 13. Jacobi's derivative formula

We now turn to quite a startling formula, which shows that theta functions give rise to modular forms in more than one way. Upto now, we have considered $\vartheta_{ij}(0,\tau)$ for $(i,j) = (0,0), (0,1)$ and $(1,0)$. Since $\vartheta_{11}(0,\tau)=0$, this gives nothing new; however, if we consider instead

$$\frac{\partial}{\partial z}\vartheta_{11}(z,\tau)\Big|_{z=0}$$

which we abbreviate as $\vartheta'_{11}(0,\tau)$, we find, e.g., that $\vartheta'_{11}(0,\tau)^2$ is a modular form (of weight 3 and level 4), etc. In fact, this is an immediate consequence of (Prop. 9.2 and) the following:

**Proposition 13.1.** For all $\tau \in H$, we have Jacobi's derivative formula, namely,

$$(J_2): \qquad \vartheta'_{11}(0,\tau) = -\pi\vartheta_{00}(0,\tau)\,\vartheta_{01}(0,\tau)\,\vartheta_{10}(0,\tau)$$

Proof. By definition, the Fourier expansion of $\vartheta'_{11}(0,\tau)$ is given by

$$\vartheta'_{11}(0,\tau) = \frac{\partial}{\partial z}\Big(\sum_{n\in\mathbb{Z}} \exp(\pi i(n+\tfrac{1}{2})^2\tau + 2\pi i(n+\tfrac{1}{2})(z+\tfrac{1}{2}))\Big)\Big|_{z=0}$$

$$= 2\pi i \sum_{n\in\mathbb{Z}} (n+\tfrac{1}{2}) \exp[\pi i(n+\tfrac{1}{2})^2\tau + \pi i(n+\tfrac{1}{2})]$$

$$= 2\pi \sum_{n\in\mathbb{Z}} (-1)^{n+1}(n+\tfrac{1}{2}) \exp(\pi i(n+\tfrac{1}{2})^2\tau).$$

In terms of the variable $q = \exp(\pi i\tau)$, the local coordinate at the cusp $i\infty$, we get:

$$\vartheta'_{11}(0,\tau) = -2\pi[q^{1/4} - 3q^{9/4} + 5q^{25/4} - \ldots\ldots\ldots].$$

So the formula $(J_2)$ reads as

$$[q^{1/4}-3q^{9/4}+5q^{25/4}-7q^{49/4}+\dots] = [1+2q+2q^4+2q^9+\dots]\cdot[1-2q+2q^4-2q^9+\dots][q^{1/4}+q^{9/4}+q^{25/4}+\dots]$$

which the reader may enjoy verifying for 3 or 4 terms (we have taken the expansions on the right form page 1.16, § 5 above). We may prove this as follows: start with the Riemann theta formula ($R_{10}$) above and expand it around the origin, getting:

$$[\vartheta_{00}+\tfrac{1}{2}\vartheta''_{00}x^2+\dots][\vartheta_{01}+\tfrac{1}{2}\vartheta''_{01}y^2+\dots][\vartheta_{10}+\tfrac{1}{2}\vartheta''_{10}u^2+\dots][\vartheta'_{11}v+\frac{\vartheta'''_{11}}{6}v^3+\dots]$$

$$+[\vartheta_{01}+\tfrac{1}{2}\vartheta''_{01}x^2+\dots][\vartheta_{00}+\tfrac{1}{2}\vartheta''_{00}y^2+\dots][\vartheta'_{11}u+\frac{\vartheta'''_{11}}{6}u^3+\dots][\vartheta_{10}+\tfrac{1}{2}\vartheta''_{10}u^2+\dots]$$

$$+[\vartheta_{10}+\tfrac{1}{2}\vartheta''_{10}x^2+\dots][\vartheta'_{11}y+\frac{\vartheta'''_{11}}{6}y^3+\dots][\vartheta_{00}+\tfrac{1}{2}\vartheta''_{00}u^2+\dots][\vartheta_{01}+\tfrac{1}{2}\vartheta''_{01}v^2+\dots]$$

$$+[\vartheta'_{11}x+\frac{\vartheta'''_{11}}{6}x^3+\dots][\vartheta_{10}+\tfrac{1}{2}\vartheta''_{10}y^2+\dots][\vartheta_{01}+\tfrac{1}{2}\vartheta''_{01}u^2+\dots][\vartheta_{00}+\tfrac{1}{2}\vartheta''_{00}v^2+\dots]$$

$$=2[\vartheta'_{11}x_1+\frac{\vartheta'''_{11}}{6}x_1^3+\dots][\vartheta_{10}+\tfrac{1}{2}\vartheta''_{10}y_1^2+\dots][\vartheta_{01}+\tfrac{1}{2}\vartheta''_{01}u_1^2+\dots][\vartheta_{00}+\tfrac{1}{2}\vartheta''_{00}v_1^2+\dots]$$

where $x_1 = \frac{1}{2}(x+y+u+v)$, $y_1 = \frac{1}{2}(x+y-u-v)$, $u_1 = \frac{1}{2}(x-y+u-v)$, $v_1 = \frac{1}{2}(x-y-u+v)$. Now comparing the coefficients of any cubic term, say $x^3$, on both sides (the result is the same for all the cubic terms), we get:

$$\frac{1}{6}\vartheta'''_{11}\vartheta_{10}\vartheta_{01}\vartheta_{00} = \frac{1}{24}\vartheta_{00}\vartheta_{01}\vartheta_{10}\vartheta'''_{11}+\frac{1}{8}\vartheta_{00}\vartheta_{01}\vartheta''_{10}\vartheta'_{11}$$
$$+\frac{1}{8}\vartheta_{00}\vartheta''_{01}\vartheta_{10}\vartheta'_{11}+\frac{1}{8}\vartheta''_{00}\vartheta_{01}\vartheta_{10}\vartheta'_{11}$$

Or, equivalently,

$$0 = \frac{\vartheta'''_{11}}{\vartheta'_{11}} - \frac{\vartheta''_{00}}{\vartheta_{00}} - \frac{\vartheta''_{01}}{\vartheta_{01}} - \frac{\vartheta''_{10}}{\vartheta_{10}}\ .$$

But in view of the Heat equation

$$\frac{\partial^2}{\partial z^2}\vartheta_{ij} = 4\pi i \frac{\partial}{\partial \tau}\vartheta_{ij}$$

the above is also equivalent to

$$0 = \frac{\partial}{\partial \tau}[\log \vartheta'_{11} - \log \vartheta_{00} - \log \vartheta_{01} - \log \vartheta_{10}],$$

or $\vartheta'_{11}/\vartheta_{00}\vartheta_{01}\vartheta_{10}$ is a constant function of $\tau$ (on H). Letting $\tau \longrightarrow i\infty$, we see that asymptotically

$\vartheta'_{11} \sim -2\pi \exp(\pi i \tau/4)$ and $\vartheta_{00}\vartheta_{01}\vartheta_{10} \sim 2\exp(\pi i \tau/4)$, hence the constant is $-\pi$. This proves the formula $(J_2)$.

As a consequence of this formula, we have:

Corollary 13.2. $\vartheta'_{11}(0,\tau)^2$ (besides being a modular form of weight 3 and level 4) is a cusp form, i.e., it vanishes at all the cusps (since at each cusp one of the 3 modular forms $\vartheta^2_{ij}(0,\tau)$, $(i,j) \neq (1,1)$, vanishes).

We shall find later a large class of differential operators which applied to theta functions give modular forms. However, only isolated generalisations of Jacobi's formula $(J_2)$ have been found and it remains a tantalising and beautiful result but not at all well-understood!

## § 14. Product expansion of $\vartheta$ and applications

We shall devote the rest of this chapter to discussing some arithmetical applications of the theory of theta functions. No one can doubt that a large part of the interest in the theory of theta functions had always been derived from its use as a powerful tool for deriving arithmetic facts. We saw this already in § 7, when we evaluated Gauss Sums along the way in proving the functional equation.

We will divide the arithmetic applications into 3 groups constituting the contents of this and the subsequent sections. The first group consists in a set of startingly elegant evaluations of infinite formal products which go back to Euler and Jacobi. Their connection with theta functions comes from the idea of expanding $\vartheta(z,\tau)$ in an infinite product. However, these product formulae are special to the one variable case. Since the zeros of $\vartheta(z,\tau)$ break up into the doubly infinite set

$$z = \tfrac{1}{2} + \tfrac{1}{2}\tau + n + m\tau;\ m,n \in \mathbb{Z},$$

it is natural to expect that $\vartheta$ will have a corresponding product decomposition.

In fact, note that

$$\exp[\pi i(2m+1)\tau - 2\pi iz] = -1 \Longleftrightarrow 2\pi iz - \pi(2m+1)\tau = (2n+1)\pi i,\ n \in \mathbb{Z}$$

$$\Longleftrightarrow z = \tfrac{1}{2}(2m+1)\tau + \tfrac{1}{2}(2n+1)\,.$$

This suggests that $\vartheta(z,\tau)$ should be of the form

$$\prod_{m\in\mathbb{Z}} (1 + \exp[\pi i(2m+1)\tau - 2\pi iz])$$

upto some nowhere vanishing function as a factor. To obtain convergence, we separate the terms with $2m+1 > 0$ & $2m+1 < 0$ and consider the infinite product:

$$p(z,\tau) = \prod_{m\in\mathbb{Z}^+} \{(1+\exp[\pi i(2m+1)\tau - 2\pi iz])(1+\exp[\pi i(2m+1)\tau + 2\pi iz])\}$$

To see that $p(z,\tau)$ converges (absolutely and uniformly on compact sets), we have only to show that the 2 series

$$\sum_{m\in\mathbb{Z}^+} \exp[\pi i(2m+1)\tau \pm 2\pi iz]$$

have the same property: in fact, if $\mathrm{Im}\, z < c$ & $\mathrm{Im}\,\tau \geq d > 0$, then

$$\left|\exp[\pi i(2m+1)\tau \pm 2\pi i z]\right| \leq (\exp 2\pi c)(\exp -\pi d)^{2m+1}$$

etc., hence $p(z, \tau)$ converges strongly. Clearly $p(z, \tau)$ has the same zeros as $\vartheta(z, \tau)$. Now we have the following:

Proposition 14.1. An infinite product expansion for $\vartheta(z, \tau)$ is given by

$$(J_3): \vartheta(z, \tau) = \prod_{m \in \mathbb{N}} (1-\exp \pi i(2m)\tau) \prod_{m \in \mathbb{Z}^+} \{(1+\exp[\pi i(2m+1)\tau - 2\pi i z]).$$
$$(1+\exp[\pi i(2m+1)\tau + 2\pi i z])\}.$$

Proof. We write the right hand side as $c(\tau) \, . \, p(z, \tau)$. Observe that the convergence of the function

$$c(\tau) = \prod_{m \in \mathbb{N}} (1-\exp \pi i(2m)\tau)$$

is immediate, and is nowhere vanishing on H. On the other hand, $p(z, \tau)$ has the same periodic behaviour in $z$ as $\vartheta$: in fact, we see that

a) $p(z+1, \tau) = p(z, \tau)$ (clear from definition of $p(z, \tau)$),

b) $$p(z,+\tau, \tau) = \prod_{m \in \mathbb{N}} (1+\exp[\pi i(2m+1)\tau - 2\pi i(z+\tau)])(1+\exp[\pi i\tau - 2\pi i(z+\tau)]).$$
$$\prod_{m \in \mathbb{Z}^+} (1+\exp[\pi i(2m+1)\tau + 2\pi i(z+\tau)])$$
$$= \prod_{m \in \mathbb{N}} (1+\exp[\pi i(2m-1)\tau - 2\pi i z])\{\exp(-\pi i\tau - 2\pi i z).$$
$$(1+\exp[\pi i\tau + 2\pi i z])\} \prod_{m \in \mathbb{Z}^+} (1+\exp[\pi i(2m+3)\tau + 2\pi i z])$$
$$= \exp(-\pi i\tau - 2\pi i z). \, p(z, \tau).$$

Therefore, we must have

$$(*) \qquad \vartheta(z, \tau) = c'(\tau)\, p(z, \tau)$$

for some (nowhere zero) holomorphic function $c'(\tau)$. To show that $c'(\tau) = c(\tau)$, we will use Jacobi's derivative formula $(J_2)$ from the previous section! In fact, substituting $z+\frac{1}{2}$, $z+\frac{1}{2}\tau$, $z+\frac{1}{2}+\frac{1}{2}\tau$ for $z$ in (*), we get:

$$\vartheta_{01}(z,\tau) = c'(\tau)\prod_{m\in\mathbb{Z}}\{(1-\exp[\pi i(2m+1)\tau-2\pi iz])(1-\exp[\pi i(2m+1)\tau+2\pi iz])\}$$

$$\vartheta_{10}(z,\tau) = c'(\tau)\,(\exp\pi i\tau/4)\,[\exp\pi iz + \exp(-\pi iz)]\,.$$
$$\prod_{m\in\mathbb{N}}\{(1+\exp[\pi i2m\tau-2\pi iz])(1+\exp[\pi i\,2m\tau+2\pi iz])\}$$

$$\vartheta_{11}(z,\tau) = ic'(\tau)\,(\exp\pi i\tau/4)\,[\exp\pi iz - \exp(-\pi iz)]\,.$$
$$\prod_{m\in\mathbb{N}}\{(1-\exp[\pi i2m\tau-2\pi iz])(1-\exp[\pi i\,2m\tau+2\pi iz])\}\,.$$

Thus we get:

$$\vartheta_{00}(0,\tau) = c'(\tau)\prod_{m\in\mathbb{Z}^+}(1+\exp\pi i\,(2m+1)\,\tau)^2$$

$$\vartheta_{01}(0,\tau) = c'(\tau)\prod_{m\in\mathbb{Z}^+}(1-\exp\pi i\,(2m+1)\,\tau)^2$$

$$\vartheta_{10}(0,\tau) = 2c'(\tau)\,(\exp\pi i\tau/4)\prod_{m\in\mathbb{N}}(1+\exp\pi i(2m)\,\tau)^2$$

$$\vartheta'_{11}(0,\tau) = -2\pi c'(\tau)\,(\exp\pi i\tau/4)\prod_{m\in\mathbb{N}}(1-\exp\pi\, i(2m)\,\tau)^2\,.$$

(The last one is obtained by writing

$$\vartheta_{11}(z,\tau) = [\exp\pi iz - \exp(-\pi iz)]f(z)$$

and noting that simply $\vartheta'_{11}(0,\tau) = 2\pi i f(0)$). Now substituting into Jacobi's formula $(J_2)$, we get:

$$-2\pi c'(\tau)(\exp \pi i\,\tau/4) \prod_{m\in\mathbb{N}} (1-\exp \pi i(2m)\tau)^2$$

$$= -2\, c'(\tau)^3 \prod_{m\in\mathbb{N}} (1+\exp \pi i(2m)\tau)^2 \prod_{m\in\mathbb{Z}^+} (1-\exp \pi i(4m+2)\tau)^2$$

or

$$c'(\tau)^2 = \left[\frac{\prod_{m\in\mathbb{N}}(1-\exp \pi i(2m)\tau)}{\prod_{m\in\mathbb{N}}(1+\exp \pi i(2m)\tau)\prod_{m\in\mathbb{Z}^+}(1-\exp \pi i(4m+2)\tau)}\right]^2$$

Now cancelling 2nd part of the denominator against the terms in the numerator corresponding to $m = 1, 3, 5, \ldots,$ we get:

$$c'(\tau)^2 = \left[\frac{\prod_{m\in\mathbb{N}}(1-\exp \pi i(4m)\tau)}{\prod_{m\in\mathbb{N}}(1+\exp \pi i(2m)\tau}\right]^2 .$$

Writing $1-\exp \pi i(4m)\,\tau = (1+\exp \pi i 2m\,\tau)(1-\exp \pi i\, 2m\,\tau)$ and cancelling gives

$$c'(\tau)^2 = \prod_{m\in\mathbb{N}} (1-\exp \pi i(2m)\tau)^2 = c(\tau)^2 .$$

But since $\lim_{\operatorname{Im}\tau\to\infty} c(\tau) = 1$, this shows that

$$c(\tau) = \prod_{m\in\mathbb{N}} (1-\exp \pi i(2m)\tau),$$

as required. This proves the formula $(J_3)$.

Some applications: In terms of the variables $q = \exp \pi i\,\tau$ and $w = \exp \pi i z$, the formula $(J_3)$ reads:

$$(P_1): \quad \sum_{m\in\mathbb{Z}} q^{m^2} w^{2m} = \prod_{m\in\mathbb{N}} (1-q^{2m}) \prod_{m\in\mathbb{Z}^+} \{(1+q^{2m+1}w^2)(1+q^{2m+1}w^{-2})\}.$$

An elementary proof of this striking identity can be found in Hardy and Wright, p. 280. Setting $w = 1$ and $w = i$ respectively give equally striking special cases:

$$(P_2): \qquad \sum_{m\in\mathbb{Z}} q^{m^2} = \prod_{m\in\mathbb{N}} (1-q^{2m}) \prod_{m\in\mathbb{Z}^+} (1+q^{2m+1})^2$$

$$(P_3): \qquad \sum_{m\in\mathbb{Z}} (-1)^m q^{m^2} = \prod_{m\in\mathbb{N}} (1-q^{2m}) \prod_{m\in\mathbb{Z}^+} (1-q^{2m+1})^2 .$$

However, the most striking variant of all arises when we look at $\vartheta_{\frac{1}{6},\frac{1}{2}}(0,3\tau)$: we have

$$\begin{aligned}\vartheta_{\frac{1}{6},\frac{1}{2}}(0,3\tau) &= (\exp \pi i/6)(\exp \pi i\tau/12)\, \vartheta_{oo}(\tfrac{1}{2}+\tfrac{1}{2}\tau, 3\tau)\\ &= (\exp \pi i/6)(\exp \pi i\tau/12) \prod_{m\in\mathbb{N}} (1-\exp \pi i(2m)\, 3\tau).\\ &\qquad \prod_{m\in\mathbb{Z}^+} \{(1-\exp[\pi i(2m+1)\, 3\tau \pm \pi i\tau])\}\\ &= (\exp \pi i/6)(\exp \pi i\tau/12) \prod_{k\in\mathbb{N}} (1-\exp \pi i(2k)\tau).\end{aligned}$$

On the other hand, we have

$$\begin{aligned}\vartheta_{\frac{1}{6},\frac{1}{2}}(0,3\tau) &= \sum_{m\in\mathbb{Z}} \exp[\pi i(m+\tfrac{1}{6})^2\, 3\tau + 2\pi i(m+\tfrac{1}{6})\tfrac{1}{2}]\\ &= (\exp \pi i/6)(\exp \pi i\tau/12) \sum_{m\in\mathbb{Z}} (-1)^m \exp \pi i(3m^2+m)\tau .\end{aligned}$$

Thus we get in terms of q:

$$(P_4): \qquad \sum_{m\in\mathbb{Z}} (-1)^m q^{3m^2+m} = \prod_{m\in\mathbb{N}} (1-q^{2m})$$

which was first proved by Euler. A final identity of the same genre is found by returning to the formula for $\vartheta'_{11}$ and substituting $c(\tau)$, we find:

$$\vartheta'_{11}(0,\tau) = -2\pi(\exp \pi i \tau/4) \prod_{m \in \mathbb{N}} (1-\exp \pi i(2m)\tau)^3 .$$

But we have (from § 13)

$$\vartheta'_{11}(0,\tau) = -\pi(\exp \pi i \tau/4) \sum_{m \in \mathbb{Z}} (-1)^m (2m+1)(\exp \pi i(m^2+m)\tau).$$

Thus we get in terms of q :

$$(P_5) : \qquad \sum_{m \in \mathbb{Z}} (-1)^m (2m+1) q^{m^2+m} = 2 \prod_{m \in \mathbb{N}} (1-q^{2m})^3 .$$

Combining $(P_4)$ and $(P_5)$, we deduce that

$$[\vartheta_{\frac{1}{6},\frac{1}{2}}(0,3\tau)]^3 = \frac{1}{2\pi i} \vartheta'_{\frac{1}{2},\frac{1}{2}}(0,\tau)$$

hence that $\vartheta_{\frac{1}{6},\frac{1}{2}}(0,3\tau)$ has value zero at all the cusps. It is the simplest $\vartheta_{a,b}$ with this property. Among higher powers,

$$[\vartheta_{\frac{1}{6},\frac{1}{2}}(0,3\tau)]^{24} = \exp(2\pi i \tau) \prod_{k \in \mathbb{N}} (1-\exp \pi i(2k)\tau)^{24}$$

is the famous <u>Δ-function of Jacobi.</u> The reader can check easily from $(P_5)$, $(J_2)$ and Table V that it is a modular form of level 1. It is the simplest modular form of level 1 vanishing at all the cusps! $(P_4)$ and $(P_5)$ are in fact the first two of an infinite sequence of evaluations of the coefficients $a_{m,k}$ in:

$$\prod_{m \in \mathbb{N}} (1-q^{2m})^k = \sum_{m \in \mathbb{N}} a_{m,k} q^m$$

discovered by I. Macdonald whenever k is the dimension of a semi-simple Lie group! (cf. M. Demazure, Identités de Macdonald, Exp. 483, <u>Séminaire Bourbaki</u>, 1975/76; Springer Lecture Notes No. 567 (1977)).

These results may perhaps be considered more combinatorial than arithmetical. They have interesting applications in the theory of the partition function $p(n)$ : we refer the reader to Hardy and Wright, An Introduction to the Theory of Numbers, Oxford University Press, 1945, Chapters 19 and 20.

## § 15. Representation of an integer as sum of squares

The most famous arithmetic application of theta series is again due to Jacobi and is this: let

$$r_k(n) = \#\{(n_1, \ldots, n_k) \in \mathbb{Z}^k / n_1^2 + \ldots + n_k^2 = n\}$$

$$= \text{number of representations of } n \text{ as a sum of } k \text{ squares}$$

(counting representations as distinct even if only the order or sign is changed). Thus, for instance, $r_2(5) = 8$ as

$$5 = 2^2 + 1^2 = 2^2 + (-1)^2 = (-2)^2 + 1^2 = (-2)^2 + (-1)^2$$
$$= 1^2 + 2^2 = (-1)^2 + 2^2 = 1^2 + (-2)^2 = (-1)^2 + (-2)^2 .$$

In terms of $q = \exp \pi i \tau$, recall that we have

$$\vartheta(0, \tau) = \sum_{n \in \mathbb{Z}} q^{n^2}$$

and hence

$$\vartheta(0, \tau)^k = \sum_{n_1 \in \mathbb{Z}} \cdots \sum_{n_k \in \mathbb{Z}} q^{n_1^2 + \ldots + n_k^2}$$
$$= \sum_{n \in \mathbb{Z}^+} r_k(n) q^n$$

i.e., $\vartheta(0, \tau)^k$ is the generating function for these coefficients $r_k(n)$. For $k = 4$, we have the following:

Theorem 15.1 (Jacobi): For $n \in \mathbb{N}$, we have

$$r_4(n) = \begin{cases} 8 \sum_{d|n} d & \text{if } n \text{ is odd} \\ 24 \sum_{d|n \,\&\, d \text{ odd}} d & \text{if } n \text{ is even.} \end{cases}$$

Proof. One way to prove this result is to deduce it from infinite product expansion of $\vartheta$, but a more significant way (the significance being in having more generalisations) is by relating $\vartheta^4$ to Eisenstein series following Hardy and Siegel(*). We proceed in four steps.

(1). Eisenstein series: the basic Eisenstein series are the holomorphic functions

$$E_k(\tau) = \sum_{\substack{m,n\in\mathbb{Z}\\(m,n)\neq(0,0)}} \frac{1}{(m\tau+n)^k} = \sum_{\substack{\lambda\in\Lambda_\tau\\ \lambda\neq 0}} \frac{1}{\lambda^k}\,.$$

Here $k$ is a positive even integer, and $k \geq 4$ to ensure absolute convergence. In fact, as the lattice points are evenly distributed, the sum

$$\sum_{0\neq\lambda\in\Lambda_\tau} |\lambda|^{-k}$$

behaves like the integral

$$\iint_{|x+iy|>1} |x+iy|^{-k}\,dx\,dy$$

which converges only if $k > 2$ (for, the integral = const. $\int_1^\infty t^{1-k}\,dt$).

Note that if $\begin{pmatrix} a & b\\ c & d\end{pmatrix} \in SL(2,\mathbb{Z})$, then:

$$\begin{aligned}
E_k\left(\frac{a\tau+b}{c\tau+d}\right) &= \sum_{(m,n)\neq(0,0)} \frac{1}{\left[m\left(\frac{a\tau+b}{c\tau+d}\right)+n\right]^k}\\
&= (c\tau+d)^k \sum_{(m,n)\neq(0,0)} \frac{1}{[(am+cn)\tau+(bm+dn)]^k}\\
&= (c\tau+d)^k\, E_k(\tau)
\end{aligned}$$

---

(*) The proof given here was explained to me by S. Raghavan, and it follows an idea of Hecke.

because the mapping from $\mathbb{Z}^2$ to $\mathbb{Z}^2$ sending

$$(m, n) \longmapsto (am + cm, bm + dn)$$

is a bijection. Moreover, its Fourier expansion is easily calculated: we group the terms as follows:

$$E_k(\tau) = \sum_{n \neq 0} \frac{1}{n^k} + \sum_{m \neq 0} \Big( \sum_{n \in \mathbb{Z}} \frac{1}{(m\tau + n)^k} \Big)$$

$$= 2 \sum_{n \in \mathbb{N}} \frac{1}{n^k} + 2 \sum_{m \in \mathbb{N}} \Big( \sum_{n \in \mathbb{Z}} \frac{1}{(m\tau+n)^k} \Big) \text{ (since k is even)}$$

$$= 2 \Big[ \zeta(k) + \sum_{m \in \mathbb{N}} \Big( \sum_{n \in \mathbb{Z}} \frac{1}{(m\tau + n)^k} \Big) \Big]$$

(where $\zeta(s)$ is the Riemann zeta function). Already the terms in the parentheses are periodic for $\tau \longmapsto \tau+1$. To expand them, start with the well-known infinite product expansion:

$$\sin \pi z = \pi z \prod_{n \in \mathbb{N}} \Big(1 - \Big(\frac{z}{n}\Big)^2\Big).$$

Taking the logarithmic derivative, we get:

$$\frac{\pi \cos \pi z}{\sin \pi z} = \frac{1}{z} + \sum_{n \in \mathbb{N}} \Big( \frac{2z}{z^2 - n^2} \Big)$$

(the series on the right converges absolutely and uniformly on compact subsets in $\mathbb{C}$). This can be rewritten in a series which converges if $\operatorname{Im} z > 0$ as:

$$-i\pi\Big(1+2 \sum_{n \in \mathbb{N}} \exp 2\pi i n z\Big) = -i\pi \frac{1+\exp 2\pi i z}{1-\exp 2\pi i z}$$

$$= \pi \frac{\cos \pi z}{\sin \pi z}$$

$$= \frac{1}{z} + \sum_{n \in \mathbb{N}} \Big( \frac{1}{z+n} + \frac{1}{z-n} \Big).$$

(The term in brackets cannot be broken up, otherwise convergence is lost).

Differentiating this (k-1) times, we get:

$$-(2\pi i)^k \sum_{n\in\mathbb{N}} n^{k-1} \exp 2\pi inz = (-1)^{k-1}(k-1)!\,[\frac{1}{z^k} + \sum_{n\in\mathbb{N}} (\frac{1}{(z+n)^k} + \frac{1}{(z-n)^k})].$$

As soon as $k \geq 2$, the term in brackets can be broken up, so we get:

$$(*)_k : \sum_{n\in\mathbb{Z}} \frac{1}{(z+n)^k} = \frac{(-2\pi i)^k}{(k-1)!} (\sum_{n\in\mathbb{N}} n^{k-1} \exp 2\pi inz).$$

Thus if $k > 2$, we get:

$$E_k(\tau) = 2\,[\zeta(k) + \sum_{m\in\mathbb{N}} (\frac{(-2\pi i)^k}{(k-1)!} \sum_{n\in\mathbb{N}} n^{k-1} \exp 2\pi in(m\tau))]$$

$$= 2[\zeta(k) + \frac{(-2\pi i)^k}{(k-1)!} (\sum_{N\in\mathbb{N}} (\sum_{n\mid N} n^{k-1}) \exp 2\pi i N\tau)]$$

$$= [\zeta(k) + \frac{(-2\pi i)^k}{(k-1)!} (\sum_{n\in\mathbb{N}} \sigma_{k-1}(n) \exp 2\pi in\tau)]$$

(where $\sigma_k(n) = \sum_{d\mid n} d^k$ = sum of $k^{th}$ powers of all positive divisors of n).
This identity still remains valid even for the case when $k = 2$ if only $E_2(\tau)$ is summed carefully, i.e., sum first over n and then over m, in which case, this calculation shows that it converges conditionally. In particular, this shows that

$$\lim_{\mathrm{Im}\,\tau \to \infty} E_k(\tau) = 2\zeta(k)$$

hence $E_k(\tau)$ has good behaviour at the cusps, and therefore if $k \geq 4$, $E_k(\tau)$ is a modular form of weight k and level 1. Note that its Fourier coefficients are the more elementary number-theoretic functions $\sigma_k(n)$. Our plan is ultimately to write $\vartheta^4(0,\tau)$ as an Eisenstein series related to $E_2(\tau)$: first notice that $\vartheta^4(0,\tau)$ is a modular form of weight 2 whereas $E_2$ does not even converge absolutely, so our proof of the functional equation for $E_2$

breaks down(*)! However, recall that $\vartheta^4$ is only a modular form for $\Gamma_{1,2}$; so what we can do is to:

(2). Modify the Eisenstein series $E_2$ slightly,

thereby loosing intentionally a bit of periodicity but gaining absolute convergence. Let

$$E_2^{\vartheta}(\tau) = \sum_{m,n\in\mathbb{Z}} \left[\frac{1}{(2m\tau+(2n+1))^2} - \frac{1}{((2m+1)\tau+2n)^2}\right].$$

Since

$$\left|\frac{1}{(2m\tau+(2n+1))^2} - \frac{1}{((2m+1)\tau+2n)^2}\right| = \left|\frac{(4m+1)\tau^2+(4n-4m)\tau-(4n+1)}{(2m\tau+2n+1)^2((2m+1)\tau+2n)^2}\right|$$

$$\leq \frac{Am+Bn}{(m^2+n^2)^2} \leq \frac{C}{(m^2+n^2)^{3/2}},$$

we get that $E_2^{\vartheta}(\tau)$ is absolutely convergent on compact sets. Moreover, if we sum over $n$ first, then both the series

$$\sum_{n\in\mathbb{Z}} \frac{1}{(2m\tau+2n+1)^2} \quad \text{and} \quad \sum_{n\in\mathbb{Z}} \frac{1}{((2m+1)\tau+2n)^2}$$

are absolutely convergent and so:

$$E_2^{\vartheta}(\tau) = \sum_{m\in\mathbb{Z}} \left[\sum_{n\in\mathbb{Z}} \frac{1}{(2m\tau+2n+1)^2} - \sum_{n\in\mathbb{Z}} \frac{1}{((2m+1)\tau+2n)^2}\right]$$

Let us now evaluate the inner sums:

---

(*) In fact, $E_2$ defined by conditional convergence, is not a modular form: cf. Weil, Elliptic functions according to Eisenstein and Kronecker, Springer, 1976.

**1st term:**

(i) if $m = 0$; $\sum_{n \in \mathbb{Z}} \frac{1}{(2n+1)^2} = 2 \prod_{p \text{ odd prime}} (1-p^{-2})^{-1} = 2(1-\frac{1}{4}) \prod_{p \text{ prime}} (1-p^{-2})^{-1}$

$$= 2(1-\frac{1}{4})\,\zeta(2) = \frac{3}{2} \cdot \frac{\pi^2}{6} = \frac{\pi^2}{4} \;.$$

(ii) if $m > 0$; $\sum_{n \in \mathbb{Z}} \frac{1}{(2m\tau + 2n+1)^2} = \frac{1}{4} \sum_{n \in \mathbb{Z}} \frac{1}{[(m\tau + \frac{1}{2}) + n]^2}$

$$= -\pi^2 \sum_{n \in \mathbb{N}} n \exp 2\pi i n(m\tau + \tfrac{1}{2}) \quad (\text{by } (*)_2)$$

$$= -\pi^2 \sum_{n \in \mathbb{N}} (-1)^n n \exp 2\pi i (nm)\tau$$

(iii) if $m < 0$; changing $m, n$ to $-m, -n-1$, we see that the same formula as above holds with $-m$ instead of $m$. In other words, for $m \neq 0$, we have

$$\sum_{n \in \mathbb{Z}} \frac{1}{(2m\tau + 2n+1)^2} = \pi^2 \sum_{n \in \mathbb{N}} (-1)^{n+1} \exp 2\pi i(|m| n)\tau$$

**2nd term:**

(i) if $2m + 1 > 0$: $\sum_{n \in \mathbb{Z}} \frac{1}{((2m+1)\tau + 2n)^2} = -\pi^2 \sum_{n \in \mathbb{N}} n \exp \pi i n (2m+1)\tau$

(ii) if $2m+1 < 0$ : changing $m, n$ to $-m-1, -n$, we find that the same formula holds with $-m-1$ instead of $m$.

As each of these can be summed over $m$ individually, we get:

$$E_2^{\vartheta}(\tau) = \frac{\pi^2}{4} - 2\pi^2 \sum_{m \in \mathbb{N}} \Big( \sum_{n \in \mathbb{N}} (-1)^n n \exp 2\pi i n m \tau\Big) + 2\pi^2 \sum_{m \in \mathbb{Z}^+} \sum_{n \in \mathbb{N}} n \exp \pi i n (2m+1)\tau$$

$$= \frac{\pi^2}{4} + 2\pi^2 \sum_{N \in \mathbb{N}} \Big( \big( \sum_{\substack{n | N \\ \frac{N}{n} \text{ even}}} (-1)^{n+1} n\big) \exp \pi i N\tau\Big) + 2\pi^2 \sum_{N \in \mathbb{N}} \Big( \sum_{\substack{n | N \\ \frac{N}{n} \text{ odd}}} n\Big) \exp \pi i N \tau$$

$$= \frac{\pi^2}{4}\{1+8 \sum_{\substack{N \in \mathbb{N} \\ N \text{ even}}} \Bigg[\sum_{\substack{n|N \\ n, \frac{N}{n} \text{ even}}} -n + \sum_{\substack{n|N \\ n \text{ odd}}} n + \sum_{\substack{n|N \\ \frac{N}{n} \text{ odd}}} n\Bigg] \exp \pi i N \tau + 8 \sum_{\substack{N \in \mathbb{N} \\ N \text{ odd}}} (\sum_{n|N} n) \exp \pi i N \tau \}$$

$$= \frac{\pi^2}{4}\{1+24 \sum_{\substack{N \in \mathbb{N} \\ N \text{ even}}} ((\sum_{\substack{n|N \\ n \text{ odd}}} n) \exp \pi i N \tau) + 8 \sum_{\substack{N \in \mathbb{N} \\ N \text{ odd}}} (\sum_{n|N} n) \exp \pi i N \tau \}$$

(where we have used the identity $2^r - 2^{r-1} - 2^{r-2} - \ldots - 2 + 1 = 3$ to conclude that

$$\sum_{d|N_1} d\, 2^r - \sum_{1 \le s \le r-1,\ d|N_1} 2^s d + \sum_{d|N_1} d = 3 \sum_{d|N_1} d \text{ for } N = 2^r N_1, r > 0 \text{ and } N_1 \text{ odd}).$$

(3) $E_2^{\vartheta}(\tau)$ is a modular form:

As an obvious consequence of the above, we have a functional equation for $E_2^{\vartheta}(\tau)$, namely,

$$E_2^{\vartheta}(\tau+2) = E_2^{\vartheta}(\tau) .$$

Moreover, we have:

$$E_2^{\vartheta}(-\frac{1}{\tau}) = \sum_{m,n \in \mathbb{Z}} \left[\frac{1}{(-\frac{2m}{\tau}+2n+1)^2} - \frac{1}{(-\frac{2m+1}{\tau}+2n)^2}\right]$$

$$= -\tau^2 \Big( \sum_{m,n \in \mathbb{Z}} \left[\frac{1}{(2n\tau-2m-1)^2} - \frac{1}{((2n+1)\tau-2m)^2}\right] \Big)$$

$$= -\tau^2 E_2^{\vartheta}(\tau+2)$$

because

$$E_2^{\vartheta}(\tau+2) = \sum_{m,n \in \mathbb{Z}} \left[\frac{1}{2m\tau+4m+2n+1)^2} - \frac{1}{((2m+1)\tau+4m+2n+2)^2}\right]$$

and now replacing m by n and n by -2n -m-1, we see that this sum is precisely the previous one.

Thus

$$E_2^\vartheta\left(-\frac{1}{\tau}\right) = -\tau^2 E_2^\vartheta(\tau)$$

This shows that $E_2^\vartheta(\tau)$ has the same functional equation as $\vartheta(0,\tau)^4$ for the subgroup $\Gamma_{1,2} \subset SL(2,\mathbb{Z})$, namely,

$$(*) \qquad E_2^\vartheta\left(\frac{a\tau+b}{c\tau+d}\right) = (-1)^c (c\tau+d)^2 E_2^\vartheta(\tau).$$

Now modulo $\Gamma_2$, there are 3 cusps, i.e.,

$\{p/q \mid p \text{ odd}, q \text{ even}\} \cup \{\infty\}$, $\{p/q \mid p, q \text{ odd}\}$ and $\{p/q \mid p \text{ even}, q \text{ odd}\}$,

or $\infty$, 1 and 0 for short. Notice that the extra substitution $\tau \longmapsto -\frac{1}{\tau}$ in $\Gamma_{1,2}$ carries 0 to $\infty$. On the other hand, by its Foruier expansion, $E_2^\vartheta$ is bounded for $\mathrm{Im}\,\tau \geq c$, so $E_2^\vartheta$ has the bound for a modular form at $\infty$ and hence at all rational points representing the cusps $\infty$ and 0. As for 1, we must expand $E_2^\vartheta(-1/\tau+1)$ (in a Fourier series) as we did for $E_2^\vartheta$ and check that its only terms are $\exp \pi i N\tau$, $N \geq 0$. But

$$E_2^\vartheta\left(-\frac{1}{\tau}+1\right) = \tau^2 \sum_{n,m \in \mathbb{Z}} \left[\frac{1}{((2n+1)\tau+2m)^2} - \frac{1}{((2n+1)\tau+2m-1)^2}\right]$$

and this can be expanded in a Foruier series just like $E_2^\vartheta(\tau)$. (In fact, since $(2n+1)\tau$ is never 0, there is no constant term either). The conclusion therefore is that $E_2^\vartheta(\tau)$ is a modular form. The final crucial step is the following:

(4). $\boxed{E_2^\vartheta(\tau) = (\pi^2/4)\,\vartheta_{00}(0,\tau)^4}$ Note that this identity at once implies the theorem by comparison of the Fourier coefficients. On the other hand, to prove (4) itself, a direct verification of Jacobi's theorem for say the first 10 coefficients $r_4(n)$, shows that $E_2^\vartheta(\tau) - (\pi^2/4)\,\vartheta_{00}(0,\tau)^4$ has a zero of order 10 at the cusp $i\infty$, i.e., that the meromorphic function

$$f = E_2^{\vartheta}/\vartheta_{00}^4 - \pi^2/4$$

on $H/\tilde{\Gamma}$ has a zero of order 10 at the cusp $i\infty$. But $\vartheta_{00}$ is zero only at the cusp 1, and there has double zero: so $f$ can have at most 8 poles which is a contradiction to # poles = # zeros unless $f \equiv 0$.

This might be considered a lazy man's way to finish this argument! In fact, there are more elegant ways to go about: viz.,

(a) it is quite easy to check that the space of modular forms $f$ for $\Gamma_{1,2}$ with the functional equation (*) in (3) above, is one-dimensional. So merely knowing that the values of $E_2^{\vartheta}$ and $(\pi^2/4)\vartheta_{00}^4$ at the cusp $i\infty$ are equal is enough to conclude that they are equal everywhere; or,

(b) one could note that $E_2^{\vartheta}$ is zero at the same cusps where $\vartheta_{00}$ is. Hence $E_2^{\vartheta}/\vartheta_{00}^4$ is bounded at all cusps, and so by Liouville's theorem it is a constant.

Most important point here is to see the underlying philosophy of modular forms: these are always finite dimensional vector spaces of functions characterised by their functional equations and behaviour at the cusps. Thus between functions arising from quite different sources which turn out to be modular forms, one can expect to find surprising identities! In particular, Jacobi's formula has been vastly generalised by Siegel. We shall describe without proof Siegle's formula in Chapter II: it shows that for any number of variables, certain weighted averages of the representation numbers $r(n)$ of $n$ by a finite set of quadratic forms can be expressed by divisor sums $\sigma_k$; or equivalently, that certain polynomials in the $\vartheta_{a,b}$ are equal to Eisenstein series.

## § 16. Theta and Zeta

The most exciting arithmetic application of modular forms, however, is one which is partly a dream at this point: a dream however that rests on several complete theories and quite a few calculations. The dream has grown from ideas of Hecke, taking clearer shape under the hands of Weil, and now has been vastly extended and analysed by Langlands.

The germ of this theory lies in the fact that the Mellin transform carries the Jacobi Theta function to the Riemann Zeta function and that in this way, the functional equation for $\vartheta$ implies the one for $\zeta$. We want to explain this and generalise it following Hecke in this section, postponing Hecke's most original ideas to the next two sections. The Mellin transform $M$ carries a function $f(x)$ defined for $x \in \mathbb{R}^{>}$ with suitable bounds at $o$ and $\infty$ to an analytic function $Mf(s)$ defined by

$$Mf(s) = \int_0^\infty f(x)\, x^s \frac{dx}{x}, \quad a < \mathrm{Re}(s) < b$$

and it is inverted by

$$f(x) = \frac{1}{2\pi i} \int_{c-i\infty}^{c+i\infty} Mf(s)\, x^{-s} ds, \quad c \in (a,b).$$

It is just the Fourier-Laplace transform in another guise because $x = \exp y$ carries $-\infty < y < \infty$ to $o < x < \infty$, and in terms of $f(\exp y)$, we have:

$$Mf(s) = \int_{-\infty}^{\infty} f(\exp y) \exp(ys)\, dy$$

which for $\mathrm{Re}\, s = 0$ is the Fourier transform of $y \longmapsto f(\exp y)$ and, with suitable bounds on $f(\exp y)$ as $y \longrightarrow \pm\infty$, $Mf(s)$ is analytic in $s = u+iv$ when $u$ is in the same interval $(a,b)$. The usual inversion gives

$$f(\exp y)\exp cy = \frac{1}{2\pi i}\int_{-\infty}^{\infty} Mf(c+iv)\exp(-ivy)\,dv$$

or

$$f(\exp y) = \frac{1}{2\pi i}\int_{c-i\infty}^{c+i\infty} Mf(s)\exp(-sy)\,ds,$$

as asserted. In particular, apply this to

$$f(x) = 2\sum_{n\in\mathbb{N}} \exp(-\pi n^2 x)\ .$$

Note then that

$$1+f(-ix) = \sum_{n\in\mathbb{Z}} \exp \pi i n^2 x = \vartheta(o,\ x)$$

or

$$f(x) = \vartheta(o,\ ix) - 1\ .$$

As we have already seen (in § 9), recall that

$$|f(x)| \leq C\ \exp(-\pi x) \qquad \text{as } x \longrightarrow \infty$$

$$|f(x)| \leq C\, x^{-\frac{1}{2}} \qquad \text{as } x \longrightarrow 0$$

Thus provided $\mathrm{Re}(s) > 1$, we have:

$$M(\vartheta(o,ix) - 1)(\tfrac{1}{2}s) = 2\int_0^{\infty}\Big(\sum_{n\in\mathbb{N}} \exp(-\pi n^2 x)\Big)\, x^{\frac{1}{2}s}\,\frac{dx}{x}\ .$$

Since this integral converges absolutely, interchanging the order of $\int$ and $\Sigma$ gives:

$$M(\qquad "\qquad)(\tfrac{1}{2}s) = 2\sum_{n\in\mathbb{N}}\Big(\int_0^{\infty} \exp(-\pi n^2 x)\, x^{\frac{1}{2}s}\,\frac{dx}{x}\Big).$$

In the $n^{th}$ integral, we make the substitution $y = \pi n^2 x$, obtaining

$$M( \quad " \quad )(\tfrac{1}{2}s) = 2 \sum_{n \in \mathbb{N}} (\pi n^2)^{-\frac{1}{2}s} \int_0^\infty \exp(-y)\, y^{\frac{1}{2}s} \frac{dy}{y}$$

$$= 2\pi^{-\frac{1}{2}s} \Big( \sum_{n \in \mathbb{N}} n^{-s} \Big) \int_0^\infty \exp(-y)\, y^{\frac{1}{2}s} \frac{dy}{y}$$

$$= 2\pi^{-\frac{1}{2}s} \zeta(s)\, \Gamma(\tfrac{1}{2}s) .$$

Thus we have the <u>fundamental formula</u> (for Re $(s) > 1$):

$$(*) \qquad 2\pi^{-\frac{1}{2}s} \zeta(s) \Gamma(\tfrac{1}{2}s) = \int_0^\infty (\vartheta(o, ix) - 1)\, x^{\frac{1}{2}s} \frac{dx}{x} .$$

This can be used to prove in one step the most well-known elementary properties of $\zeta(s)$:

<u>Proposition 16.1.</u> The Riemann zeta function $\zeta(s) = \sum_{n \in \mathbb{N}} n^{-s}$, $\mathrm{Re}(s) > 1$, has a meromorphic continuation to the whole s-plane with a simple pole at $s = 1$ and satisfies a functional equation, namely,

$$\xi(s) = \xi(1-s) \text{ where } \xi(s) = \pi^{-\frac{1}{2}s} \zeta(s)\, \Gamma(\tfrac{1}{2}s).$$

<u>Proof.</u> Recall that we have

$$\vartheta(o, i/y) = y^{\frac{1}{2}} \vartheta(o, iy), \; y \in \mathbb{R}^{>} .$$

We use this in (*) above as follows:

$$2\pi^{-\frac{1}{2}s} \zeta(s)\, \Gamma(\tfrac{1}{2}s) = \int_1^\infty (\vartheta(o, ix) - 1)\, x^{\frac{1}{2}s} \frac{dx}{x} + \int_0^1 (\vartheta(o, ix) - 1)\, x^{\frac{1}{2}s} \frac{dx}{x} .$$

The first integral converges for all $s \in \mathbb{C}$ and defines an entire function. As for the second:

$$\int_0^1 (\vartheta(o, ix) - 1)\, x^{\frac{1}{2}s} \frac{dx}{x} = \Big( \int_0^1 \frac{\vartheta(o, i/x)}{x^{\frac{1}{2}}} x^{\frac{1}{2}s} \frac{dx}{x}\Big) - \int_0^1 x^{\frac{1}{2}s-1} dx$$

$$= \Big( \int_1^\infty \vartheta(o, iy)\, y^{\frac{1}{2}(1-s)} \frac{dy}{y}\Big) - \frac{2}{s}\,;\ (y = \frac{1}{x})$$

$$= \Big( \int_1^\infty (\vartheta(o, iy) - 1)\, y^{\frac{1}{2}(1-s)} \frac{dy}{y} + \int_1^\infty y^{-\frac{1}{2}(1+s)} dy\Big) - \frac{2}{s}$$

$$= \Big( \int_1^\infty \vartheta(o, iy) - 1)\, y^{\frac{1}{2}(1-s)} \frac{dy}{y}\Big) - \frac{2}{1-s} - \frac{2}{s}$$

$$= \text{entire function } - \frac{2}{1-s} - \frac{2}{s}\,.$$

Thus

$$\xi(s) = \pi^{-\frac{1}{2}s} \zeta(s)\Gamma(\tfrac{1}{2}s) = -\frac{1}{s} - \frac{1}{1-s} + \frac{1}{2}\int_1^\infty (\vartheta(o, ix) - 1)(x^{\frac{1}{2}s} + x^{\frac{1}{2}(1-s)})\frac{dx}{x}$$

which shows that $\xi(s)$ is meromorphic with simple poles at $s = 0$ and 1, and is unchanged for the substitution $s \longmapsto 1-s$. Recall that $\Gamma(s)$ has a pole at $s = 0$ and hence $\zeta(s)$ is analytic at $s = 0$. This completes the proof of the proposition.

Dirichlet series: The above considerations can be generalised as follows: fix (for simplicity) an integral weight $k \geq 0$ and a level $n \geq 1$, and suppose that $f(z)$, $z \in H$, is a modular form of weight $k$ and level $n$. (N.B: we are replacing the usual variable $\tau$ by $z$ here). Then $f(z)$ can be expanded: first we have

$$f(z) = \sum_{m \in \mathbb{Z}^+} a_m \exp(2\pi i m z/n)$$

because $\begin{pmatrix} 1 & n \\ 0 & 1 \end{pmatrix} \in \Gamma_n$ and hence $f(z+n) = f(z)$, etc. We associate to $f$, the formal Dirichlet series, defined by

$$Z_f(s) = \sum_{m \in \mathbb{N}} a_m m^{-s}\,.$$

Then we have the following:

Theorem 16.2(Hecke): The Dirichlet series $Z_f(s)$ converges(*) if $\mathrm{Re}\,(s) > k+1$ and has a meromorphic continuation to the whole s-plane with one simple pole at $s = k$. Moreover, there is a decomposition of $\mathrm{Mod}_k^{(n)} = V_{+1} \oplus V_{-1}$ such that whenever $f \in V_\epsilon$ ($\epsilon = \pm 1$), $Z_f(s)$ satisfies a functional equation, namely

$$\left(\frac{n}{2\pi}\right)^s \Gamma(s) Z_f(s) = \epsilon \left(\frac{n}{2\pi}\right)^{k-s} \Gamma(k-s)\, Z_f(k-s).$$

Proof. Let us first find bounds for the growth of the coefficients $a_m$:

Lemma 16.3. There exists a constant $C$ such that

$$|a_m| \leq C\, m^k, \quad \forall m \in \mathbb{N}.$$

Proof. Let us evaluate the $m^{th}$ Fourier coefficient of $f$ by integrating $f$ along the line $z = x+i/m$, $0 \leq x \leq n$. We have

$$\int_0^n f(x+i/m) \exp(-2\pi imx/n)dx$$

$$= \sum_{\ell \in \mathbb{Z}^+} \frac{a_\ell}{n} \int_0^n \exp\left[\frac{2\pi i \ell(x+i/m) - 2\pi imx}{n}\right] dx$$

$$= n\ a_m \exp(-2\pi/n).$$

On the other hand, by Remark 9.4, we have

$$|f(x+i/m)| \leq C_o \,.\, m^k \quad \text{for} \quad m \geq 2 \ ( > 2/3^{\frac{1}{2}})$$

for some constant $C_o$. Thus

$$|a_m| \leq \frac{1}{n} \exp(2\pi/n) \int_0^n |f(x+i/m)| dx$$

$$\leq \exp(2\pi/n) \,.\, C_o \,.\, m^k$$

$$\leq C \,.\, m^k, \quad \forall m \geq 1,$$

(*) In fact, it converges if $\mathrm{Re}(s) > k$ but we won't prove this.

as asserted.

Proof of Theorem 16.2. The convergence of $Z_f(s)$ is immediate for $\mathrm{Re}(s) > k+1$ (by Lemma 16.3). Now we relate $f$ and $Z_f$ by the Mellin transform as before:

$$M(f(ix) - a_o)(s) = \int_0^\infty \Big(\sum_{m\in\mathbb{N}} a_m \exp(-2\pi mx/n)\Big) . x^s \frac{dx}{x}$$

$$= \sum_{m\in\mathbb{N}} a_m \int_0^\infty \exp(-2\pi mx/n)\, x^s \frac{dx}{x} .$$

Replacing $x$ by $y = 2\pi mx/n$ in the $m^{th}$ integral, we find:

$$M(f(ix) - a_o)(s) = \sum_{m\in\mathbb{N}} \int_0^\infty \exp(-y) \Big(\frac{n}{2\pi m}\Big)^s y^s \frac{dy}{y}$$

$$= \Big(\frac{n}{2\pi}\Big)^s \Big(\sum_{m\in\mathbb{N}} a_m m^{-s}\Big) \int_0^\infty \exp(-y)\, y^s \frac{dy}{y}$$

$$= \Big(\frac{n}{2\pi}\Big)^s Z_f(s)\, \Gamma(s) .$$

(Here we are assuming $\mathrm{Re}(s) > k+1$ and we can interchange the order of $\Sigma$ and $\int$ because the calculation shows that

$$\sum_{m\in\mathbb{N}} |a_m| \int_0^\infty \exp(-2\pi mx/n)\, x^{\mathrm{Re}(s)} \frac{dx}{x} < \infty) .$$

Now since $f(z) \in \mathrm{Mod}_k^{(n)}$, we see that $g(z) \in \mathrm{Mod}_k^{(n)}$ where

$$g(z) = \Big(\frac{z}{i}\Big)^{-k} f\Big(-\frac{1}{z}\Big)$$

$$= \sum_{m\in\mathbb{Z}^+} b_m \exp(2\pi imz/n), \text{ say.}$$

But then we have:

$$\Big(\frac{n}{2\pi}\Big)^s Z_f(s)\, \Gamma(s) = M(f(ix) - a_o)(s)$$

$$= \int_1^\infty (f(ix) - a_o)\, x^s \frac{dx}{x} + \int_0^1 (f(ix) - a_o) x^s \frac{dx}{x}$$

$$= \int_1^\infty (f(ix) - a_o)\, x^s \frac{dx}{x} - \frac{a_o}{s} + \int_0^1 x^{-k} g(ix)\, x^s \frac{dx}{x}$$

$$= \int_1^\infty (\text{"})\, x^s \frac{dx}{x} - \frac{a_o}{s} + \frac{b_o}{s-k} + \int_1^\infty (g(iy) - b_o)\, y^{k-s} \frac{dy}{y};\ (y = \frac{1}{x}).$$

This shows that $Z_f(s)$ has a meromorphic continuation as asserted. Since $\Gamma(s)$ has a simple pole at $s = 0$, $Z_f(s)$ has a simple pole only at $s = k$. Finally, the map $f \longmapsto g$ obviously defines an automorphism of $\mathrm{Mod}_k^{(n)}$ of order 2, so decompose $\mathrm{Mod}_k^{(n)}$ into the 2 eigen spaces

$$V_{+1} = \{ g = f \} \text{ and } V_{-1} = \{ g = -f \}$$

and the above identity gives immediately the stated functional equation. This completes the proof of the theorem.

Examples of Dirichlet series. What sort of Dirichlet series do we get as functions $Z_f(s)$? Here are 2 simple cases:

Example 1. (Epstein Zeta function): Let

$$f(z) = \vartheta(o,z)^{2k} = \sum_{n_1,\dots n_{2k} \in \mathbb{Z}} \exp\, (\pi i (n_1^2 + \dots + n_{2k}^2)\, z$$

$$= \sum_{m \in \mathbb{Z}^+} r_{2k}(m) \exp \pi imz\ .$$

We know that $f(z)$ has a functional equation for $\Gamma_{1,2}$ including $z \longmapsto z+2$ and $z \longmapsto -\frac{1}{z}$; and upto a root of unity in its functional equation, it is a modular form of weight $k$ and level 2. The associated Dirichlet series is

$$Z_{\vartheta^{2k}}(s) = \sum_{m \in \mathbb{N}} r_{2k}(m)\, m^{-s} = \sum_{\substack{n_1,\dots,n_{2k} \in \mathbb{Z} \\ (n_1,\dots,n_{2k}) \neq (0,\dots,0)}} \frac{1}{(n_1^2 + \dots + n_{2k}^2)^s}$$

This is the simplest Epstein zeta function. It is a particular case of the zeta functions

$$\sum_{0 \neq \lambda \in \Lambda} \frac{1}{Q(\lambda)^s}$$

where $Q$ is a positive definite quadratic form and $\Lambda$ is a lattice.

It follows from what we have proved that $Z_{\vartheta, 2k}(s)$ is a meromorphic function with a simple pole at $s = k$, and has a functional equation for the substitution $s \longmapsto k-d$.

Example 2. (Dirichlet series associated to Eisenstein series): Let

$$f(z) = E_k(z) = \sum_{\substack{m,n \in \mathbb{Z} \\ (m,n) \neq (0,0)}} (mz+n)^{-k}, \quad k \in 2\mathbb{N}$$

Recall that $E_k$ is an Eisenstein series introduced in the previous section. According to the calculations made there of its Fourier expansion, we have

$$Z_{E_k}(s) = c_k \sum_{m \in \mathbb{N}} \sigma_{k-1}(m)\, m^{-s} \quad \text{where} \quad \sigma_t(m) = \sum_{d|m} d^t, \; c_k = 2\frac{(-2\pi i)^k}{(k-1)!}\,.$$

On the other hand, we have

$$\begin{aligned} \zeta(s)\zeta(s-k+1) &= \sum_{m,n \in \mathbb{N}} m^{-s}\, n^{-s+k-1} \\ &= \sum_{m,n \in \mathbb{N}} n^{k-1} \cdot (mn)^{-s} \\ &= \sum_{\ell \in \mathbb{N}} \sigma_{k-1}(\ell)\, \ell^{-s}. \end{aligned}$$

Thus

$$Z_{E_k}(s) = c_k \cdot \zeta(s)\, \zeta(s-k+1)\,.$$

We can now state the central theme of the dream referred to at the beginning of this section: it is to say that the class of Dirichlet series that arise naturally in arithmetic (viz., Artin's L-series attached to finite representations of $Gal(\bar{\mathbb{Q}}/\mathbb{Q})$, Hasse-Weil $\zeta$-functions attached to algebraic varieties over $\mathbb{Q}$ by considering their points mod p, and generalisations thereof(*) ) is the same as the class of modular form Dirichlet series $Z_f$ plus their generalisations associated to certain types of modular forms(**)! Now by their very definition, every arithmetic Dirichlet series is equal to an Euler product:

$$Z(s) = \prod_{\text{primes } p} (\text{rational function of } p^{-s}).$$

A pre-requisite for the coincidence of the 2 classes is that $Z_f(s)$ has an Euler product for a set of modular forms f spanning $Mod_k^{(n)}$: this is the main point of Hecke's further ideas that we now turn to (in the last 2 sections of this chapter).

---

(*) cf. J.-P. Serre, Zeta and L-functions, Arithmetical Algebraic Geometry: Proc. of a conference held at Purdue University (1963), Harper & Row, Publishers, New York, 1965.

(**) cf. A. Borel, Formes automorphes et séries de Dirichlet, Séminaire Bourbaki, 1974/75, Exp. 466; Springer Lecture Notes No. 514, 1976. As Serre has explained to me, for the Dirichlet series $Z_f$ to be part of this dream, one wants to put some restriction on the eigenvalues of the invariant differential operators acting on these forms, e.g. most of Maass' non-holomorphic forms are not included.

## § 17. Hurwitz maps

Let us describe abstractly the group-theoretic background to theta functions and modular forms:

(a) for a complex torus, we have:

(i) lattice $\Lambda$ acting on $\mathbb{C}$ by translation

(ii) the orbit space $E = \mathbb{C}/\Lambda$

(iii) functions on $\mathbb{C}$, such as $\vartheta(z)$, automorphic for the action of $\Lambda$, i.e., periodic upto a factor $e_\lambda(z)$, $\lambda \in \Lambda$.

The basic idea in unwinding the function theory of $E$ is to shrink the lattice $\Lambda$ to $\ell\Lambda$, vary the function $\vartheta(z)$ to the functions $\vartheta_{a,b}(z)$, giving an interplay of the group-theory and the function-theory. A key fact here is that when the automorphic equation

$$(*) \qquad f(z+\lambda) = e_\lambda(z)\, f(z)$$

is required only for $\lambda \in \ell\Lambda$, then $g(z) = f(z+\mu)$ for $\mu$ in the larger lattice $\frac{1}{\ell}\Lambda$ again satisfies (*) for all $\lambda \in \ell\Lambda$. Shrinking $\Lambda$ further, eventually translations with respect to all points in $\mathbb{Q}.\Lambda$ are incorporated in the function-theory.

(b) for modular forms, we have:

(i) $SL(2,\mathbb{Z})$ acting on $H$

(ii) the orbit space $H/SL(2,\mathbb{Z})$

(iii) modular forms on $H$.

As before, we can replace $SL(2,\mathbb{Z})$ by the smaller groups $\Gamma_n$. Then in place of the one Riemann surface $H/SL(2,\mathbb{Z})$, we obtain a whole tower of Riemann surfaces, namely:

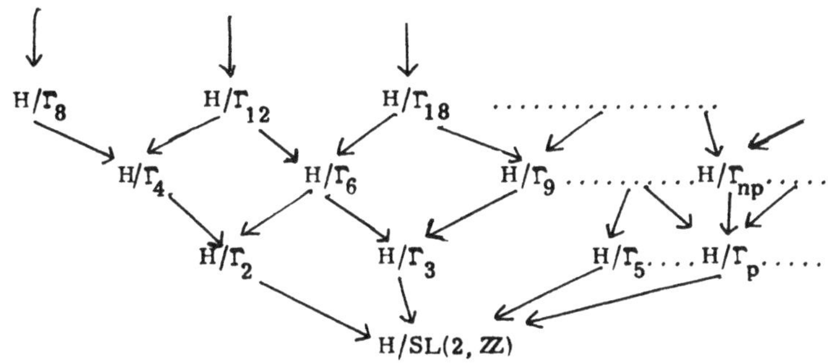


The group $SL(2,\mathbb{Z})/\Gamma_n = SL(2,\mathbb{Z}/n\mathbb{Z})$ acts on $H/\Gamma_n$ and on the basic function-theoretic entity, namely, the ring $\mathrm{Mod}^{(n)}$ of modular forms of level n.

However, notice a difference here: we are not enlarging the group $SL(2,\mathbb{Z})$. Just as $\Lambda \subset \mathbb{Q}.\Lambda$, also $SL(2,\mathbb{Z}) \subset SL(2,\mathbb{Q})$. Actually it is better to think of $SL(2,\mathbb{Z})$ as $GL(2,\mathbb{Z})_+$, the elements of $GL(2,\mathbb{Z})$ with positive determinant. Then $SL(2,\mathbb{Z}) \subset GL(2,\mathbb{Q})_+$ is a bigger enlargement of "integral" by "rational" elements. To incorporate $GL(2,\mathbb{Q})_+$ into the picture; note that $\gamma \in GL(2,\mathbb{Q})_+$ does not map any $H/\Gamma_n$ to itself unless $\gamma\Gamma_n\gamma^{-1} = \Gamma_n$ which occurs only in the trivial cases, i.e., $\gamma \binom{a\ 0}{0\ a} \in SL(2,\mathbb{Z})$, $a \in \mathbb{Q}$. Instead, what occurs is that $\gamma\Gamma_n\gamma^{-1} \subset \Gamma_m$ if $n = (ad-bc)\, m$ where

$$\gamma\begin{pmatrix} k & 0 \\ 0 & k \end{pmatrix} = \begin{pmatrix} a & b \\ c & d \end{pmatrix};\ a,b,c,d \in \mathbb{Z},\ (a,b,c,d) = 1.$$

We therefore get a new map which we call a Hurwitz map:

$$T_\gamma : H/\Gamma_n \xrightarrow{\approx} H/\gamma\Gamma_n\gamma^{-1} \xrightarrow[\text{(canonical covering)}]{} H/\Gamma_m$$

$$z \longmapsto \gamma z$$

acting "sideways" on our tower. In view of the elementary divisor theorem, the new maps are all compositions of $SL(2,\mathbb{Z})/\Gamma_n$ acting on $H/\Gamma_n$ and the basic maps

$$T_\ell = T_\gamma : H/\Gamma_{\ell n} \longrightarrow H/\Gamma_n$$

given by translation by $\gamma = \begin{pmatrix} \ell & 0 \\ 0 & 1 \end{pmatrix}$, i.e., $\tau \longmapsto \ell\tau$.

Hurwitz studied these in the form of the "<u>modular correspondences</u>", i.e., we have 2 maps:

$$H/\Gamma_\ell \begin{array}{l} \xrightarrow{\tau \mapsto \ell\tau} H/\Gamma_1 \\ \xrightarrow{\text{nat. proj.}} H/\Gamma_1 \end{array}$$

hence we can consider the image

$$H/\Gamma_\ell \longrightarrow C_\ell \subset (H/\Gamma_1) \times (H/\Gamma_1)$$

Clearly $C_\ell$ is just the image in $(H \times H)/(\Gamma_1 \times \Gamma_1)$ of the locus of points $(\tau, \ell\tau)$ in $H \times H$. It is called the $\ell^{\text{th}}$ <u>modular correspondence.</u> If $H/\Gamma_1$ is taken as the moduli of complex tori, then it is easy to check the following:

<u>Proposition 17.1.</u> Let $\tau_1, \tau_2 \in H$. Then

$$(E_{\tau_1}, E_{\tau_2}) \in C_\ell \Longleftrightarrow \left\{ \begin{array}{l} \exists \text{ a covering map } \pi : E_{\tau_1} \longrightarrow E_{\tau_2} \text{ whose covering} \\ \text{group is translations on } E_{\tau_1} \text{ by a cyclic group of order } \ell \end{array} \right\}$$

Currently, the most fashionable approach to this new structure is to consider the inverse limit

$$\mathcal{H} = \varprojlim_n H/\Gamma_n$$

of all the Riemann surfaces in the tower. This $\mathcal{H}$ is <u>not</u> the same as $H$ just as the real line $\mathbb{R}$ is <u>not</u> the same as the compact abelian topological group (Solenoid)

$$\varprojlim_n \mathbb{R}/n\mathbb{Z}$$

i.e., the induced map $H \longrightarrow \mathcal{H}$ is <u>not</u> even bijective!

The important point is that the Hurwitz map

$$T_\gamma : H/\Gamma_{n(ad-bc)} \longrightarrow H/\Gamma_n$$

between different spaces passes up through the tower and induces a bijective map $T_\gamma$ of $\mathcal{H}$ to itself. Thus $GL(2,\mathbb{Q})_+$ acts on the space $\mathcal{H}$.

<u>Appendix: Structure of the inverse limit $\mathcal{H}$.</u>

(1) Firstly, $\mathcal{H}$ has a kind of algebraic structure. In fact, $H/\Gamma_n$ is canonically an affine algebraic curve: abstractly this is because we can compactify it by adding a finite set of cusps, and a compact Riemann surface has a unique algebraic structure on it. Concretely, if $n = 4m$, we consider $R_n$ the ring of holomorphic functions

$$f/(\vartheta_{00}\vartheta_{01}\vartheta_{10})^{2k}, \quad f \in \mathrm{Mod}_k^{(n)}$$

which can be characterised as the $\Gamma_n$-invariant functions with "finite order poles" at the cusps. Then $H/\Gamma_n$ is the maximal ideal space of $R_n$. If $n = \ell m$, then $R_m, R_\ell \subset R_n$. Let

$$\mathcal{R} = \bigcup_n R_n = \left\{ \begin{array}{l} \text{holomorphic functions } f \text{ on } H \text{ invariant for some } \Gamma_n \text{ and} \\ \text{which have finite order poles at the cusps.} \end{array} \right\}$$

Then $\mathcal{H}$ is isomorphic to the maximal ideal space of $\mathcal{R}$, i.e., $\mathcal{H}$ is the <u>scheme</u> $\mathrm{Spec}\,\mathcal{R}$ - {generic point}, i.e., $\mathrm{Spec}\,\mathcal{R}$ minus its unique non-closed point corresponding the prime but not maximal ideal (0). (Notice that $\mathcal{R}$ is a "non-Noetherian Dedekind domain").

(2) Adélic interpretation: let us first recall the concept of adéles: the ring of rational adéles $\mathbb{A}$ is by definition

$$\mathbb{A} = \left\{ \begin{array}{l} \text{the subring of the product } \mathbb{R} \times \prod_{p \text{ prime}} \mathbb{Q}_p \text{ of elements} \\ (x_\infty ; x_2, x_3, \ldots, x_p, \ldots) \text{ such that } x_p \in \mathbb{Z}_p \text{ for all} \\ \text{but a finite number of primes } p \end{array} \right\}$$

where $\mathbb{Z}_p$ is the ring of p-adic integers in the field $\mathbb{Q}_p$ of p-adic numbers. $\mathbb{A}$ is a topological ring, if a basis of open neighbourhoods of 0 is given by

$$U_{\epsilon\{n(p)\}} = \{(x_\infty; \ldots, x_p, \ldots) \mid |x_\infty| < \epsilon, x_p \in p^{n(p)}\mathbb{Z}_p, \forall p\}$$

for various $\epsilon > 0$ and sequences $\{n(p)\}$ of non-negative integers such that $n(p) = 0$ for all but a finite number of p's. We embed $\mathbb{Q}$ in $\mathbb{A}$ diagonally, i.e., as the subring of adéles $(x_\infty ; \ldots, x_p, \ldots)$ such that $x_\infty = x_p = a/b \in \mathbb{Q}$ for all p. This makes $\mathbb{Q}$ into a discrete subgroup of $\mathbb{A}$ because if $a/b \in \mathbb{Q}$ and $|a/b|$ is small, then some non-trivial prime occurs in the denominator, so p-adically $a/b \notin \mathbb{Z}_p$.

The adéles frequently arise in studying inverse limits. The simplest case is the solenoid mentioned above:

Proposition 17.2. There is an isomorphism of topological groups:

$$(*) : \qquad \varprojlim_{n \in \mathbb{N}} \mathbb{R}/n\mathbb{Z} \approx \mathbb{A}/\mathbb{Q}$$

Proof. Let $n = \prod_p p^{n(p)}$ be the prime decomposition of $n \in \mathbb{N}$. Define a subgroup $K(n)$ of $\mathbb{A}$ by

$$K(n) = \{(0 ; \ldots, x_p, \ldots) / x_p \in p^{n(p)}\mathbb{Z}_p, \forall p\}.$$

Then $K(n)$ is compact and $\bigcap_n K(n) = (0)$, hence

$$\mathbb{A} \approx \varprojlim_n \mathbb{A}/K(n).$$

Therefore,

$$\mathbb{A}/\mathbb{Q} \approx \varprojlim_n \mathbb{A}/(\mathbb{Q} + K(n))$$

Now (*) will follow once we verify that the map

$$(*)_n : \qquad \mathbb{R}/n\mathbb{Z} \longrightarrow \mathbb{A}/(\mathbb{Q} + K(n))$$

given by

$$x \longmapsto (x; \ldots, 0, \ldots)$$

is an isomorphism. This map makes sense and is injective because

$$(n; \ldots, 0, \ldots) = (n; \ldots, n, \ldots) + (0; \ldots, -n, \ldots) \in \mathbb{Q} + K(n).$$

Surjectivity follows immediately from:

<u>Lemma 17.3</u> (Approximation for $\mathbb{Q}$) : Given a finite set $S$ of primes, integers $n(p) \geq 0$ and p-adic numbers $a_p \in \mathbb{Q}_p$ for $p \in S$; there exists a rational number $a \in \mathbb{Q}$ such that

(i) $a - a_p \in p^{n(p)} \mathbb{Z}_p, \forall p \in S$

(ii) $a \in \mathbb{Z}_p$ for all $p \notin S$.

The reader may enjoy checking this.

This example should serve as motivation for the more complicated adelic interpretation of $\mathcal{H}$. For this we consider

$$\mathcal{H}' = GL(2, \mathbb{Q}) \backslash GL(2, \mathbb{A}) / K_\infty \cdot Z_\infty$$

where $K_\infty$ and $Z_\infty$ are the subgroups of $GL(2,\mathbb{A})$ of matrices $X = (X_\infty; \ldots, X_p, \ldots)$ given by

$$X \in K_\infty \Longleftrightarrow X_\infty = \begin{pmatrix} \cos\theta, & \sin\theta \\ -\sin\theta, & \cos\theta \end{pmatrix} \text{ and } X_p = I_2, \forall p$$

and

$$X \in Z_\infty \Longleftrightarrow X_\infty = \begin{pmatrix} \lambda_\infty & 0 \\ 0 & \lambda_\infty \end{pmatrix},\ \lambda_\infty \in \mathbb{R}^* \text{ and } X_p = I_2, \forall p\,.$$

Note that determinant gives a map

$$\det : \mathcal{H}' \longrightarrow \mathbb{Q}^* \backslash \mathbb{A}^* / \mathbb{R}^> = \mathbb{A}^* / \mathbb{Q}^*.\mathbb{R}^>\,.$$

Using the unique factorisation of a fraction $a/b \in \mathbb{Q}^*$, namely,

$$a/b = (\pm 1) \prod_{p \text{ prime}} p^{n(p)}$$

where $n(p) \in \mathbb{Z}$ and $n(p) = 0$ for all but a finite set of primes, it is easy to see that

$$\mathbb{A}^* / \mathbb{Q}^*.\mathbb{R}^> \approx \prod_{p \text{ prime}} \mathbb{Z}_p^*$$

which is a compact totally disconnected space. Now we see that the connected components of $\mathcal{H}'$ are contained in $\det^{-1}(a)$, $a \in \mathbb{A}^*/\mathbb{Q}^*.\mathbb{R}^>$. Define

$$GL(2,\mathbb{A})^0 = \{ X \in GL(2,\mathbb{A}) \mid \det X \in \mathbb{Q}^*.\mathbb{R}^> \}$$

$$\mathcal{H}'_0 = GL(2,\mathbb{Q}) \backslash GL(2,\mathbb{A})^0 / K_\infty.Z_\infty = \det^{-1}(1)$$

Then $\mathcal{H}'_0$ is in fact connected as a corollary of:

Theorem 17.4. $\mathcal{H} \approx \mathcal{H}'_0$ and in this isomorphism $T_\gamma : \mathcal{H}'_0 \longrightarrow \mathcal{H}'_0$ becomes right multiplication by

$$A_\gamma = (I_2; \gamma, \ldots, \gamma, \ldots)$$

(i.e., identity on the infinite factor and right multiplication by $\gamma$ on the finite factors).

Proof. An easy generalisation of the proof of Prop. 17.2: we need the

Lemma 17.5 (Strong approximation for $SL(2,\mathbb{Q})$): Given a finite set $S$ of primes $p$, integers $n(p)$ and matrices $X_p \in SL(2, \mathbb{Q}_p)$, $p \in S$; there exists an $X \in SL(2,\mathbb{Q})$ such that

(i) $$X = \begin{pmatrix} 1+p^{n(p)}a_p, & p^{n(p)}b_p \\ p^{n(p)}c_p, & 1+p^{n(p)}d_p \end{pmatrix} . X_p \text{ for suitable } a_p, b_p, c_p, d_p \in \mathbb{Z}_p, \forall p \in S$$

(ii) $X \in SL(2, \mathbb{Z}_p)$ for all $p \notin S$.

(For a proof see Lemma 6.15 in Shimura's : Introduction to the arithmetic theory of automorphic functions, Tokyo-Princeton, 1971).

We now analyse $\mathcal{H}'_0$ in a series of steps:

Step I: The natural map

(*) $$SL(2,\mathbb{Q})\backslash SL(2,\mathbb{A})/K_\infty \longrightarrow GL(2,\mathbb{Q})\backslash GL(2,\mathbb{A})^0/K_\infty . Z_\infty$$

is an isomorphism.

It is clearly surjective because modulo suitable elements $\begin{pmatrix} a/b & 0 \\ 0 & 1 \end{pmatrix}$ in $GL(2,\mathbb{Q})$ and $\begin{pmatrix} \lambda_\infty & 0 \\ 0 & \lambda_\infty \end{pmatrix}$ in $Z_\infty$, we can alter any $X$ in $GL(2,\mathbb{A})^0$ until its determinant is 1. To see injectivity, say $X_1, X_2 \in SL(2,\mathbb{A})$ have the same images, i.e.,

$$X_1 = AX_2BC \text{ for suitable } A \in GL(2,\mathbb{Q}),\ B \in K_\infty,\ C \in Z_\infty .$$

But then we get:

$$1 = \det A \cdot \det C.$$

On the other hand, we know that $\det A \in \mathbb{Q}^*$, $\det C \in \mathbb{R}^>$ but these two have nothing common in $\mathbb{A}^*$. So $\det A = 1$, i.e., $A \in SL(2, \mathbb{Q})$, hence $\det C = 1$, i.e.,

$$C = (\pm \begin{pmatrix} 1 & 0 \\ 0 & 1 \end{pmatrix}; \ldots, I_2, \ldots)$$

Thus $C \in K_\infty$. This means that $X_1$ and $X_2$ define the same element in $SL(2,\mathbb{Q})\backslash SL(2,\mathbb{A})/K_\infty$, as required.

Step II: For $n \in \mathbb{N}$, let $n = \prod_p p^{r(p)}$. Define a subgroup $K(n)$ of $SL(2,\mathbb{A})$ by

$$K(n) = \{ (I_2; \ldots, X_p, \ldots)/X_p \in SL(2, \mathbb{Z}_p) \ \forall p \quad \text{and}$$

$$X_p = \begin{pmatrix} 1+p^{r(p)}a_p, & p^{r(p)}b_p \\ p^{r(p)}c_p, & 1+p^{r(p)}d_p \end{pmatrix}, \ a_p, b_p, c_p, d_p \in \mathbb{Z}_p \}.$$

Then it is easy to check that $K(n)$ is a compact subgroup of $SL(2,\mathbb{A})$ and that

$$\bigcap_{n \in \mathbb{N}} K(n) = \{1\}.$$

Consequently, using (as before) the fact that $SL(2,\mathbb{A}) = \varprojlim_n SL(2,\mathbb{A})/K(n)$, we get:

$$(**): \qquad SL(2,\mathbb{Q})\backslash SL(2,\mathbb{A})/K_\infty \xrightarrow{\approx} \varprojlim_n SL(2,\mathbb{Q})\backslash SL(2,\mathbb{A})/K_\infty . K(n).$$

Step III, The natural map

$$(*'): \qquad \Gamma_n\backslash SL(2,\mathbb{R})/K_\infty \longrightarrow SL(2,\mathbb{Q})\backslash SL(2,\mathbb{A})/K_\infty . K(n)$$

induced by the natural inclusion $SL(2,\mathbb{R}) \hookrightarrow SL(2,\mathbb{A})$ given by $X \longmapsto (X; \ldots, I_2, \ldots)$ is an isomorphism. To see this: let us write $SL(2,\mathbb{A}_f) = \{X \in SL(2,\mathbb{A}) \mid X_\infty = I_2\}$, i.e., the $2 \times 2$ matrices formed from the "finite" adeles. Then strong approximation (Lemma 17.5) says that

$$SL(2,\mathbb{A}_f) = SL(2,\mathbb{Q}).\ K(n)$$

hence the map in (*') above is surjective. As for its injectivity: let $X_1, X_2 \in SL(2,\mathbb{R})$ have the same image, then for some $A \in SL(2,\mathbb{Q})$, $B \in K_\infty$, $C \in K(n)$, we have

$$(X_1;\dots,I_2,\dots) = A(X_2;\dots,I_2,\dots).\ (B_\infty;\dots,I_2,\dots).(I_2;\dots,C_p,\dots)$$

$$= (A\,X_2B_\infty;\dots,A\,C_p,\dots).$$

Therefore $\forall p$, we have $AC_p = I_2$, hence $A = C_p^{-1} \in SL(2,\mathbb{Z}_p)$ and $A \equiv I_2 (\text{mod } p^{r(p)})$ where $n = p^{r(p)} n_o$ with $p \nmid n_o$. Thus $A \in \bigcap_p SL(2,\mathbb{Z}_p) = SL(2,\mathbb{Z})$ and hence $A \in \Gamma_n$. Since $X_1 = AX_2 B_\infty$, $X_1$ and $X_2$ define the same element in $\Gamma_n \backslash SL(2,\mathbb{R})/K_\infty$ as required. Finally:

Step IV. $H \approx SL(2,\mathbb{R})/K_\infty$ because $SL(2,\mathbb{R})$ acts transitively on $H$ and $K_\infty$ is the stabiliser of $i \in H$. Thus we get:

$$\begin{aligned}\mathcal{H} = \varprojlim_n \Gamma_n \backslash H &\approx \varprojlim_n \Gamma_n \backslash SL(2,\mathbb{R})/K_\infty \\ &\approx \varprojlim_n \ SL(2,\mathbb{Q})\backslash SL(2,\mathbb{A})/K_\infty . K(n) \qquad \text{(by (*''))} \\ &\approx SL(2,\mathbb{Q})\backslash SL(2,\mathbb{A})/K_\infty \quad \text{(by (**))} \\ &\approx GL(2,\mathbb{Q})\backslash GL(2,\mathbb{A})^o/K_\infty . Z_\infty \quad \text{(by (*))} \\ &= \mathcal{H}'_o ,\end{aligned}$$

as required.

Lastly, to check the action of $T_\gamma$ on $\mathcal{H}'_o$: it is clear that it suffices to check for the basic ones $T_\ell$, i.e., when $\gamma = \begin{pmatrix} \ell & 0 \\ 0 & 1 \end{pmatrix}$, $\ell \in \mathbb{N}$. Now look at the right translation on $\mathcal{H}'_o$ defined by $\begin{pmatrix} 1 & 0 \\ 0 & \ell \end{pmatrix}$, i.e.,

$$(X_\infty;\ldots,X_p,\ldots)\longmapsto(X_\infty;\ldots,X_p\begin{pmatrix}1&0\\0&\ell\end{pmatrix},\ldots)$$

This is the same as

$$(X_\infty;\ldots,X_p,\ldots)\longmapsto\left(\begin{pmatrix}1&0\\0&\ell^{-1}\end{pmatrix}X_\infty\begin{pmatrix}\ell^{\frac{1}{2}}&0\\0&\ell^{\frac{1}{2}}\end{pmatrix};\ldots,\begin{pmatrix}1&0\\0&\ell^{-1}\end{pmatrix}X_p\begin{pmatrix}1&0\\0&\ell\end{pmatrix},\ldots\right)$$

which in this form carries $SL(2,\mathbb{A})$ to itself. Restricting this to $SL(2,\mathbb{R})$, the action is

$$(X_\infty;\ldots,I_2,\ldots)\longrightarrow\left(\begin{pmatrix}\ell^{\frac{1}{2}}&0\\0&\ell^{-\frac{1}{2}}\end{pmatrix}X_\infty;\ldots,I_2,\ldots\right)$$

and acting on H this is the map

$$\tau\longmapsto\ell\tau,\quad \tau=X_\infty(i)$$

defining $T_\ell$.

This comples the proof of the theorem. We give a 3rd interpretation of $\mathcal{H}$:

(3). $\mathcal{H}$ as a moduli space: we state the result (without proof):

$$\mathcal{H}\approx\left\{\begin{array}{l}\text{Isomorphism classes of triples } (V,L,\varphi) \text{ where}\\ V=\text{one dimensional complex vector space}\\ L=\text{Lattice in } V\\ \varphi=\text{an isomorphism}: \frac{\mathbb{Q}.L}{L}\xrightarrow{\approx}\left(\frac{\mathbb{Q}}{\mathbb{Z}}\right)^2 \text{ of "determinant 1"}\end{array}\right\}$$

To explain "determinant 1" : note that the complex structure on V orients V and enables us to distinguish "orientation preserving" bases of L, i.e., those bases $e_1, e_2$ of L such that if $ie_1=ae_1+be_2$, then $b>0$. Any two such bases $e_1,e_2$ and $f_1,f_2$ are related by

$$f_1=ae_1+be_2\ \&\ f_2=ce_1+de_2$$

with $ad-bc=1$. Any such basis gives us an isomorphism

$$\varphi_o : \frac{\mathbb{Q}.L}{L} \xrightarrow{\approx} (\frac{\mathbb{Q}}{\mathbb{Z}})^2 .$$

Now $\varphi \circ \varphi_o^{-1}$ is given by a $2 \times 2$ matrix $\binom{x\ y}{z\ t}$ with entries in $\mathrm{Hom}(\mathbb{Q}/\mathbb{Z}, \mathbb{Q}/\mathbb{Z})$ (which is well-known to be $\mathbb{A}_f$, the ring of finite adèles). The requirement is that $xt-yz = 1$. In this model of $\mathcal{H}$, $T_\ell$ is the map $(V, L, \varphi) \longrightarrow (V, L', \varphi')$ where $L' \subset \mathbb{Q}.L'$ is given as the inverse image of $(\frac{1}{\ell}\mathbb{Z}/\mathbb{Z} \times (0))$ in $(\mathbb{Q}/\mathbb{Z})^2$ under the map

$$\mathbb{Q}.L \xrightarrow{\text{nat. map}} \mathbb{Q}.L/L \xrightarrow{\varphi} (\mathbb{Q}/\mathbb{Z})^2$$

and $\varphi'$ is given by

$$\begin{array}{ccc} \mathbb{Q}.L/L & \xrightarrow{\varphi} & (\mathbb{Q}/\mathbb{Z})^2 \\ \downarrow & & \downarrow \gamma = \begin{pmatrix} \ell & 0 \\ 0 & 1 \end{pmatrix} . \\ \mathbb{Q}.L'/L' & \xrightarrow{\varphi} & (\mathbb{Q}/\mathbb{Z})^2 \end{array}$$

## § 18. Hecke operators

We shall now study the action of the Hurwitz maps defined in the previous section on functions. The simplest way to define this action is to consider

$$\mathrm{Mod}_k = \bigcup_{n \in \mathbb{N}} \mathrm{Mod}_k^{(n)} .$$

These are modular forms of indefinite level: the ratio of any two functions here is a function on $\mathcal{H} = \varprojlim_n H/\Gamma_n$. Now $GL(2, \mathbb{Q})_+$ acts on this vector space by

$$f \longmapsto f^\gamma \text{ where } f^\gamma(\tau) = (c\tau+d)^{-k} f(\frac{a\tau+b}{c\tau+d}) .$$

Thus, in the limit, we jump up from having an action merely of a quotient of $SL(2, \mathbb{Z})$, to having one of $GL(2, \mathbb{Q})_+$. This action has been much studied of late in the context of the decomposition of $GL(2,\mathbb{A})$ acting on $L^2(GL(2,\mathbb{Q})\backslash GL(2,\mathbb{A}))$. Hecke looked however at the reflection of this group action on $\mathrm{Mod}_k^{(n)}$ : take any $\gamma \in GL(2., \mathbb{Q})_+$ , any $n \in \mathbb{N}$ ; then decompose $\Gamma_n \gamma \Gamma_n$ into disjoint $\Gamma_n$ left cosets:

$$(*) \qquad \Gamma_n \gamma \Gamma_n = (\Gamma_n \gamma_1) \cup (\Gamma_n \gamma_2) \cup \ldots \cup (\Gamma \gamma_t), \ \gamma_j \in GL(2, \mathbb{Q})_+ .$$

Lemma 18.1. The number $t$ of the cosets in (*) above is finite.

Proof. Let $\ell \in \mathbb{N}$ be such that

$$\begin{pmatrix} \ell & 0 \\ 0 & \ell \end{pmatrix} \gamma = \begin{pmatrix} a & b \\ c & d \end{pmatrix}, \ a, b, c, d \in \mathbb{Z},$$

and let $m = ad-bc$. Then we know that $\gamma \Gamma_{nm} \gamma^{-1} \subset \Gamma_n$ , hence $\gamma \Gamma_{nm} \subset \Gamma_n \gamma$ . So if $\Gamma_n = \bigcup_{1 \le j \le r} \Gamma_{nm} \delta_j, \delta_j \in \Gamma_n$ , then we have:

$$\Gamma_n \gamma \Gamma_n = \bigcup_{1 \le j \le r} \Gamma_n (\gamma \Gamma_{nm}) \delta_j \subset \bigcup_{1 \le j \le r} \Gamma_n \gamma \delta_j$$

and this proves the lemma.

Lemma 18.2. Let $\gamma, \gamma_j$ be as in (*) above. For $f \in \mathrm{Mod}_k^{(n)}$ , let $T_\gamma^*(f) = \sum_{1 \le j \le t} f^{\gamma_j}$. Then $T_\gamma^*(f) \in \mathrm{Mod}_k^{(n)}$ (i.e., $T_\gamma^*$ is a map of $\mathrm{Mod}_k^{(n)}$ to itself, called the Hecke operator associated to $\gamma \in GL(2,\mathbb{Q})_+$).

Proof. We have only to check the $\Gamma_n$-invariance of $T_\gamma^*(f)$: so let $\delta \in \Gamma_n$, then from (*) above, we have:

$$\bigcup_{1 \le j \le t} \Gamma_n \gamma_j = \Gamma_n \gamma \Gamma_n = \Gamma_n \gamma \Gamma_n \delta = \bigcup_{1 \le j \le t} \Gamma_n \gamma_j \delta ,$$

hence for some permutation $\sigma$ on $\{1, \ldots, t\}$ and for some $\delta_i \in \Gamma_n$ , we get:

$$\gamma_j \delta = \delta_i \gamma_{\sigma(j)} .$$

But then we get:

$$(T^*_\gamma(f))^\delta = (\sum_j f^{\gamma_j})^\delta = \sum_j f^{\gamma_j\delta} = \sum_j f^{\delta_j\gamma_{\sigma(j)}} = \sum_j f^{\gamma_{\sigma(j)}} = T^*_\gamma(f),$$

as required.

This procedure is best illustrated by the following basic example:

Lemma 18.3. Let p be a prime. Then the following 3 sets are the same:

$$\left[X \in GL(2,\mathbb{Q})/X \text{ integral \& } \det X = p\right] \text{ the double coset } SL(2,\mathbb{Z})\begin{pmatrix}1&0\\0&p\end{pmatrix}SL(2,\mathbb{Z})$$

$$= \left[SL(2,\mathbb{Z})\begin{pmatrix}p&0\\0&1\end{pmatrix} \bigcup_{0\le j\le p-1} SL(2,\mathbb{Z})\begin{pmatrix}1&j\\0&p\end{pmatrix}\right] (= \text{union of } p+1 \text{ left cosets})$$

Proof. Let $GL(2,\mathbb{Q})$ act on row vectors $(a,b)$ by right multiplication. Then an integral $X$ with $\det X = p$, carries $\mathbb{Z}^2$ onto a sub-lattice $L \subset \mathbb{Z}^2$ of index p. Moreover, $\mathbb{Z}^2.X_1 = \mathbb{Z}^2.X_2$ if and only if $\mathbb{Z}^2 = \mathbb{Z}^2.(X_1X_2^{-1})$ or $X_1X_2^{-1} \in GL(2,\mathbb{Z})$. Since $\det(X_1X_2^{-1}) = \det X_1/\det X_2 = 1, X_1X_2^{-1} \in SL(2,\mathbb{Z})$. Thus we have an isomorphism

$$SL(2,\mathbb{Z})\backslash\{X \in GL(2,\mathbb{Q})/X \text{ integral \& } \det X = p\} \cong \{\text{sub-lattices } L \subset \mathbb{Z}^2 \text{ of index } p\}.$$

But such an $L$ necessarily contains $p\mathbb{Z}^2$, so it is determined by a 1-dimensional subspace $\bar{L}$ of $(\mathbb{Z}/p\mathbb{Z})^2$. There are $p+1$ of these and the corresponding L's are the span of

$$\{(1,0),(0,p)\}, \{(1,1),(0,p)\}, \ldots, \{(1,p-1),(0,p)\} \ \& \ \{(p,0),(0,1)\}.$$

These arise from the X's respectively given by

$$\begin{pmatrix}1&0\\0&p\end{pmatrix}, \begin{pmatrix}1&1\\0&p\end{pmatrix}, \ldots, \begin{pmatrix}1&p-1\\0&p\end{pmatrix} \ \& \ \begin{pmatrix}p&0\\0&1\end{pmatrix}.$$

This proves the equality of the 1st and 3rd sets above. On the other hand, for any $L \subset \mathbb{Z}^2$ of index p, $\exists Y_1 \in SL(2,\mathbb{Z})$ such that $L.Y_1$ = span of $\{(1,0),(0,p)\}$

and hence, if $X \in GL(2,\mathbb{Q})$ is such that $L = \mathbb{Z}^2.X$, then we get:

$$\mathbb{Z}^2.X = \mathbb{Z}^2\begin{pmatrix}1 & 0\\ 0 & p\end{pmatrix}.Y_1^{-1} \quad \text{or} \quad X = Y_2\begin{pmatrix}1 & 0\\ 0 & p\end{pmatrix} Y_1^{-1} \text{ for some } Y_2 \in SL(2,\mathbb{Z}).$$

Thus the 1st and 2nd sets are equal. This proves the lemma.

This lemma enables us to explicitly compute the Hecke operator $T^*_{\begin{pmatrix}1 & 0\\ 0 & p\end{pmatrix}}$ ($= T^*_p$ for short) : let $f \in \mathrm{Mod}_k^{(1)}$, i.e., $f$ is a modular form for the full group $SL(2,\mathbb{Z})$, and let the Fourier expansion of $f(\tau)$ be given by

$$f(\tau) = \sum_{n \in \mathbb{Z}^+} a_n \exp(2\pi i n \tau).$$

Now by definition of $T^*_p$ (cf. Lemma 18.2), we have:

$$(T^*_p f)(\tau) = f^{\begin{pmatrix}p & 0\\ 0 & 1\end{pmatrix}}(\tau) + \sum_{0 \le j \le p-1} f^{\begin{pmatrix}1 & j\\ 0 & p\end{pmatrix}}(\tau)$$

$$= f(p\tau) + p^{-k}[f(\tau/p) + f((\tau+1)/p) + \dots + f((\tau+p-1)/p)]$$

$$= \sum_{n \in \mathbb{Z}^+} a_n \exp(2\pi i n p\tau) + [\sum_{n \in \mathbb{Z}^+} a_n p^{-k}. \sum_{0 \le j \le p-1} \exp(2\pi i n j/p). \exp(2\pi i n \tau)]$$

$$= \sum_{\substack{n \in \mathbb{Z}^+\\ p|n}} a_{\frac{n}{p}} \exp(2\pi i n\tau) + p^{1-k}(\sum_{n \in \mathbb{Z}^+} a_{pn} \exp 2\pi i n \tau)$$

$$= \sum_{n \in \mathbb{Z}^+} b_n \exp(2\pi i n \tau)$$

where

(*):
$$b_n = \begin{cases} p^{1-k}. a_{pn} & \text{if } p \nmid n \\ p^{1-k} a_{pn} + a_{\frac{n}{p}} & \text{if } p \mid n \end{cases}$$

An immediate consequence of this formula is the:

<u>Corollary 18.4.</u> For all primes $p_1, p_2$, the operators $T^*_{p_1}$ and $T^*_{p_2}$ commute.

Indeed, if

$$(T^*_{p_2}(T^*_{p_1} f))(\tau) = \sum_{n \in \mathbb{Z}^+} c_n \exp(2\pi i n \tau)$$

then we have

$$c_n = \begin{cases} p_2^{1-k} b_{p_2 n} & \text{if } p_2 \nmid n \\ p_2^{1-k} b_{p_2 n} + b_{\frac{n}{p_2}} & \text{if } p_2 \mid n \end{cases}$$

$$= \begin{cases} p_2^{1-k} p_1^{1-k} a_{p_1 p_2 n} & \text{if } p_2 \nmid n \text{ and } p_1 \nmid n \\ p_2^{1-k} (p_1^{1-k} a_{p_2 p_1 n} + a_{p_2 \frac{n}{p_1}}) & \text{if } p_2 \nmid n,\ p_1 \mid n \\ p_2^{1-k} (p_1^{1-k} a_{p_1 p_2 n}) + p_1^{1-k} a_{p_1 \frac{n}{p_2}} & \text{if } p_2 \mid n,\ p_1 \nmid n \\ p_2^{1-k}(p_1^{1-k} a_{p_1 p_2 n} + a_{p_2 \frac{n}{p_1}}) + p_1^{1-k} a_{p_1 \frac{n}{p_2}} + a_{\frac{n}{p_1 p_2}} & \text{if } p_1 p_2 \mid n \end{cases}$$

= symmetric in $p_1$ and $p_2$, as required.

Therefore, we can expect to find simultaneous eigenfunctions f for all $T^*_p$. In fact, suppose that

$$T^*_p f = p^{1-k} \alpha_p f, \forall p$$

Then substituting in our formula for $T^*_p f$, we find:

$$\alpha_p a_n = \begin{cases} a_{pn} & \text{if } p \nmid n \\ a_{pn} + p^{k-1} a_{\frac{n}{p}} & \text{if } p \mid n \end{cases}$$

Clearly, these formulae enable us to solve recursively for all $a_n$ as a polynomial in the $\alpha_p$'s times $a_1$. They are best solved by going over to the Dirichlet series

$$Z_f(s) = \sum_{n \in \mathbb{N}} a_n n^{-s} .$$

More precisely, summarising the discussion above, we have:

<u>Proposition 18.5.</u> (<u>Hecke</u>): Let $f \in \mathrm{Mod}_k^{(1)}$ with its Fourier expansion

$$f(\tau) = \sum_{n \in \mathbb{Z}^+} a_n \exp(2\pi i n \tau).$$

Then the following statements are equivalent:

(1) f is a simultaneous eigen function of eigen value $\alpha_p . p^{-(k-1)}$ for the Hecke operators $T_p^*$, p prime,

(2) for all $n \in \mathbb{N}$, we have

$$\alpha_p a_n = \begin{cases} a_{pn} & \text{if } p \nmid n \\ a_{pn} + p^{k-1} a_{\frac{n}{p}} & \text{if } p \mid n \end{cases}$$

(3) the associated Dirichlet series has an Euler product expansion, namely,

$$\sum_{n \in \mathbb{N}} a_n n^{-s} = Z_f(s) = a_1 . \prod_{p \text{ prime}} (1 - \alpha_p p^{-s} + p^{k-1} . p^{-2s})^{-1} .$$

(In particular, for such an f, the Fourier coefficients $a_n$, $n > 0$, are determined completely by $a_1$).

<u>Proof.</u> We have seen that (1) $\Longrightarrow$ (2) and it is a straightforward verification to see that (3) $\Longrightarrow$ (1). Assuming (2), (3) follows once we show that for any prime p and any $q \in \mathbb{N}$, $p \nmid q$, we have:

$$(1 - \alpha_p p^{-s} + p^{k-1} . p^{-2s})( \sum_{n \in \mathbb{N}, (q,n)=1} a_n n^{-s}) = \sum_{n \in \mathbb{N}, (pq,n)=1} a_n n^{-s} .$$

Let us calculate the expression on the left hand side, i.e.,

$$\sum_{(q,n)=1} a_n n^{-s} - \sum_{(q,n)=1} \alpha_p a_n (pn)^{-s} + \sum_{(q,n)=1} p^{k-1} a_n (p^2 n)^{-s}$$

$$= \sum_{(q,n)=1} a_n n^{-s} - \sum_{\substack{(q,n)=1 \\ p\nmid n}} \alpha_p a_n (pn)^{-s} - \sum_{\substack{(q,n)=1 \\ p\mid n}} \alpha_p a_n (pn)^{-s} + \sum_{\substack{m=np \\ (q,m)=1}} p^{k-1} a_{\frac{m}{p}} (pm)^{-s}$$

$$= \sum_{(q,n)=1} a_n n^{-s} - \sum_{\substack{(q,n)=1 \\ p\nmid n}} a_{pn} (pn)^{-s} - \sum_{\substack{(q,n)=1 \\ p\mid n}} (a_{pn} + p^{k-1} a_{\frac{n}{p}})(pn)^{-s} + \sum_{\substack{(q,m)=1 \\ p\mid m}} p^{k-1} a_{\frac{m}{p}} (pm)^{-s} \quad \text{(by (2))}$$

$$= \sum_{(q,n)=1} a_n n^{-s} - \sum_{(q,n)=1} a_{pn} (pn)^{-s}$$

$$= \sum_{(pq,n)=1} a_n n^{-s}, \text{ as required.}$$

From this it follows that

$$\left[ \prod_{p \text{ prime}} (1 - \alpha_p p^{-s} + p^{k-1} . p^{-2s}) \right] Z_f(s) = a_1,$$

as asserted. This completes the proof.

We do not want to develop Hecke's theory at greater length but only to give a few examples and to state his main result and the dramatic conjecture that has been made in this connection. For full proofs and details, cf. A. Ogg, Modular forms and Dirichlet series, Benjamin, 1969. To state Hecke's main result, we need some more notation: let

$$\pi: SL(2, \mathbb{Z}) \dashrightarrow SL(2, \mathbb{Z}/n\mathbb{Z})$$

be the natural map and let

$$\Gamma_n^{(1)} = \pi^{-1} \{\text{Diagonal matrices in } SL(2, \mathbb{Z}/n\mathbb{Z})\}.$$

We have

$$\Gamma_n = \ker \pi \subset \Gamma_n^{(1)} \text{ and } \Gamma_n^{(1)}/\Gamma_n = \text{diag } SL(2,\mathbb{Z}/n\mathbb{Z}) \approx (\mathbb{Z}/n\mathbb{Z})^*$$

i.e., $\Gamma_n^{(1)}/\Gamma_n$ is a finite abelian subgroup of $SL(2,\mathbb{Z})/\Gamma_n$ and acts on $\mathrm{Mod}_k^{(n)}$ by $f \longmapsto R_a(f) = f^\gamma$ where $\pi(\gamma) = a\, I_2$, $a \in (\mathbb{Z}/n\mathbb{Z})^*$. Hecke's main result is:

Theorem 18.6 (Hecke): Consider the operators on $\mathrm{Mod}_k^{(n)}$:

a) $T_\gamma^*$, $\gamma = \begin{pmatrix} a & b \\ c & d \end{pmatrix}$, $a,b,c,d \in \mathbb{Z}$, $ad-bc > 0$ and $\gcd(ad-bc, n) = 1$,

b) $f \longmapsto f^\delta$ where $\delta \in SL(2,\mathbb{Z})$ and $\delta \equiv \begin{pmatrix} a^{-1} & 0 \\ 0 & a \end{pmatrix} \pmod n$, $a \in (\mathbb{Z}/n\mathbb{Z})^*$.

Then

(1) all these operator commute,

(2) $\mathrm{Mod}_k^{(n)}$ has a basis $\{f_\alpha\}$ of simultaneous eigen functions for all of them, and

(3) the Dirichlet series $Z_{f_\alpha}$ has an Euler product:

$$Z_{f_\alpha}(s) = \begin{pmatrix} \text{rational function} \\ \text{of } p^{-s}, p \mid n \end{pmatrix} \cdot \prod_{p \nmid n} (1 - \alpha_p p^{-s} + \epsilon(p) p^{k-1} \cdot p^{-2s})^{-1}$$

where

$$\alpha_p p^{-(k-1)} = \text{eigen value of } T_p^* = T^*_{\begin{pmatrix} 1 & 0 \\ 0 & p \end{pmatrix}}$$

and

$$\epsilon(p) = \text{eigen value of } R_{(p \bmod n)}$$

What are these $f_\alpha$'s? The prime example is the Eisenstein series: we saw in § 16 that

$$Z_{E_k}(s) = c_k \zeta(s)\zeta(s-k+1) \text{ where } c_k = 2\frac{(-2\pi i)^k}{(k-1)!}$$

$$= c_k \prod_{p \text{ prime}} ((1-p^{-s})^{-1} \cdot (1-p^{k-1} \cdot p^{-s})^{-1})$$

$$= c_k \prod_{p \text{ prime}} (1 - (1+p^{k-1}) p^{-s} + p^{k-1} \cdot p^{-2s})^{-1}.$$

Hence, by Prop. 18.5, it follows that

$$T_p^*(E_k) = (1+p^{1-k})\, E_k = (1+p^{k-1})\, p^{-(k-1)}\, E_k \,.$$

In fact, it can be shown that the $f_\alpha$'s in the theorem break up into 2 disjoint groups:

(1) cusp forms and

(2) generalised Eisenstein series

$$f_\alpha(\tau) = \sum_{\substack{m_1, m_2 \in \mathbb{Z} \\ (m_1, m_2) \neq (0,0)}} \frac{c(m_1, m_2)}{(m_1\tau + m_2)^k}$$

where the $c(m_1, m_2)$'s depend only on the $m_i \pmod n$. The latter forms $f_\alpha$ have Dirichlet series $Z_{f_\alpha}(s)$ of the type

$$Z_{f_\alpha}(s) = L(\chi_1, s) \,.\, L(\chi_2, s-k+1)$$

where $L$ is the Dirichlet $L$-series (cf. Ogg for details). In particular, we get a direct sum decomposition

$$\mathrm{Mod}_k^{(n)} = (\text{Cusp forms}) \oplus (\text{Generalised Eisenstein series}).$$

We can use the results of §§ 13-16 to fit $\vartheta_{00}^4$ into this picture. In fact, we saw in § 15 that

$$Z_{\vartheta_{00}^4}(s) = 8 \sum_{\substack{n \in \mathbb{N} \\ n \text{ odd}}} \Big( \sum_{d|n} d \Big) n^{-s} + 24 \sum_{\substack{n \in \mathbb{N} \\ n \text{ even}}} \Big( \sum_{\substack{d|n \\ d \text{ odd}}} d \Big) n^{-s}$$

$$= 8 \sum_{n \in \mathbb{N}} \Big( \sum_{d|n} d \Big) n^{-s} - 32 \sum_{n \in \mathbb{N}} \Big( \sum_{d|n} d \Big) (4n)^{-s}$$

using the easily verified fact that if $n$ is even

$$3 \sum_{\substack{d|n \\ d \text{ odd}}} d = \begin{cases} \sum_{d|n} d & \text{if } 4 \nmid n \\ \sum_{d|n} d - 4 \sum_{d|\frac{n}{4}} d & \text{if } 4 | n \end{cases} .$$

Thus

$$Z_{\vartheta_{oo}^4}(s) = 8(1-4^{1-s}) \sum_{n \in \mathbb{N}} \Big( \sum_{d|n} d \Big) n^{-s}$$

$$= 8(1-4^{1-s}) \zeta(s) \cdot \zeta(1-s)$$

and hence, by Theorem 18.6', we get:

$$T_p^*(\vartheta_{oo}^4) = (1+p^{-1}) \vartheta_{oo}^4 .$$

Unfortunately, the generalisations of Jacobi's formula to other powers $\vartheta_{oo}^{2k}$ are in fact not so simple; e.g., S. Ramanujan guessed (cf. his collected works, paper 18) and Rankin proved (Am. J. Math., 1965) that:

$$\vartheta_{oo}^{2k} \in \left\{ \begin{matrix} \text{subspace spanned by the} \\ \text{Eisenstein series} \end{matrix} \right\} \Longleftrightarrow k \leq 4 .$$

The identification of the eigen functions $f_\alpha$ of the Hecke operators as polynomials in the functions $\tau \longmapsto \vartheta_{a,b}(0, n\tau)$ seems to be quite hard to describe except in the case of weight $k = 1$ or $2$. For example, take the case:

$$\mathrm{Mod}_k^{(4)} = \left\{ \begin{matrix} \text{space of homogeneous polynomials of} \\ \text{degree } k \text{ in } \vartheta_{oo}^2, \vartheta_{01}^2, \vartheta_{10}^2 \text{ modulo} \\ \text{multiples of } \vartheta_{oo}^4 - \vartheta_{01}^4 - \vartheta_{10}^4 \end{matrix} \right\}$$

Then in this space, the subspace of cusp forms is the set of multiples of $\vartheta_{oo}^2 \vartheta_{01}^2 \vartheta_{10}^2$. But it seems hard to describe in any reasonably explicit and elementary way the subspace of Eisenstein series, let alone the set of eigen functions $f_\alpha$. (cf. B. Schoeneberg, Bemerkungen zu du Eisensteinchen Reihen und ihren Anwendungen in der Arithmetik, Abh. Math. Seminar Univ.

Hamburg, Vol. 47 (1978), 201-209; for some calculations for small n). In the case of degree 1 or 2, i.e., polynomials in the $\vartheta_{a,b}(0, n\tau)$ of degree 2 or 4, the eigen functions $f_\alpha$ can be more or less found - modulo a knowledge of the arithmetic of suitable quadratic number fields, respectively quaternion algebras. This is because the theory of factorisation in imaginary quadratic fields K and in certain quaternion algebras D allows one to prove Euler products for suitable Dirichlet series

$$\sum_{\alpha \in M} \chi(\alpha) \cdot \mathrm{Nm}(\alpha)^{-s}$$

where M is a free $\mathbb{Z}$-module of rank 2 in K, respectively of rank 4 in D, and $\chi$ is a multiplicative character, just as one does for the usual Dirichlet L-series

$$\sum_{n \in \mathbb{Z}} \chi(n)\, n^{-s} .$$

But $\mathrm{Nm}(\alpha)$ is a quadratic form in 2 or 4 variables in these cases, and so there are Epstein zeta functions. Taking the inverse Mellin transform, we can express these as polynomials in $\vartheta_{a,b}(0, n\tau)$ of degree 2 or 4. For the case of quaternions where everything depends on the so called "Brandt matrices", cf. M. Eichler, The basis problem for modular forms, Springer Lecture Notes No. 320(1973).

In connection with Hecke's theorem, we want to conclude by describing a daring conjecture which arose from the work of Weil, Serre and Langlands, asserting which Dirichlet series arise from modular forms. Their conjecture is this:

Conjecture. Let K be a number field, $K_\lambda$ its $\mathscr{P}$-adic completions. Consider the continuous representations

$$\rho_\lambda : \mathrm{Gal}(\overline{\mathbb{Q}}/\mathbb{Q}) \dashrightarrow GL(2, K_\lambda).$$

Suppose for almost all $\lambda$, a $\rho_\lambda$ is given. We say that the $\rho_\lambda$'s are <u>compatible</u> if there is a finite set $S$ of rational primes $p$ such that

a) if $\lambda$ lies over a rational prime $\ell$, $\rho_\lambda$ is unramified outside $S \cup \{\ell\}$, i.e., $\rho_\lambda$ is trivial on the $p^{th}$ inertia group $I_p \subset \mathrm{Gal}(\bar{\mathbb{Q}}/\mathbb{Q})$, hence $\rho_\lambda(F_p)$, $F_p$ the $p^{th}$ Frobenius element, is well-defined and

b) for all $\lambda_1, \lambda_2$ lying over $\ell_1, \ell_2$ and all $p \notin S \cup \{\ell_1, \ell_2\}$,

$$\mathrm{Tr}\, \rho_{\lambda_1}(F_p) = \mathrm{Tr}\, \rho_{\lambda_2}(F_p)$$

$$\det \rho_{\lambda_1}(F_p) = \det \rho_{\lambda_2}(F_p)$$

and these traces and determinants are integers in $K$. We say that $\rho_\lambda$ is <u>odd</u> if $\rho_\lambda(c)$ is conjugate to $\begin{pmatrix} 1 & 0 \\ 0 & -1 \end{pmatrix}$ where $c \in \mathrm{Gal}(\bar{\mathbb{Q}}/\mathbb{Q})$ is complex conjugation. Such compatible families of representations (at least in $GL(n)$, some $n$) arise in great abundance from the theory of étale cohomology of algebraic varieties. For any such family, we can form the Dirichlet series:

$$Z^o_{\{\rho_\lambda\}}(s) = \prod_{p \notin S} (1 - \mathrm{Tr}_{\rho_\lambda}(F_p)p^{-s} + \det_{\rho_\lambda}(F_p)p^{-2s})^{-1}.$$

We may be able to supply suitable p-factors for $p \in S$. Now building on partial results of Kuga, Sato and Shimura, Deligne proved the following:

<u>Theorem.</u> Let $f \in \mathrm{Mod}_k^{(n)}$ and suppose that $f_\alpha$ is an eigen function for the Hecke operators $T_p^*$, $p \nmid n$. Let $Z^o_{f_\alpha}$ be the product of the p-factors in $Z_{f_\alpha}$ for $p \nmid n$. Then there exists a compatible family $\{\rho_\lambda\}$ of odd 2-dimensional representations with $S = \{$primes dividing $n\}$, such that

$$Z^o_{f_\alpha}(s) = Z^o_{\{\rho_\lambda\}}(s).$$

The case $k = 1$ is analysed in Deligne-Serre, Formes Modulaires de poids 1, Annales de Sci. Ecole Norm. Sup., t. 7(1974), 507 - : precisely in this case, all the $\rho_\lambda$ 's coincide and come from one

$$\rho : \mathrm{Gal}(\overline{\mathbb{Q}}/\mathbb{Q}) \dashrightarrow GL(2,K)$$

with finite image. The conjecture in question is the converse to this theorem, i.e., every series $Z^0_{\{\rho_\lambda\}}(s)$ is $Z^0_{f_\alpha}(s-m)$ for some $\alpha$, $m \geq 0$ !

## References and Questions

For many of the topics treated, especially for several treatments of the functional equation (§7) an easy place to read more is:

R. Bellman, A Brief Introduction to Theta Functions, Holt, 1961.

For a systematic treatment of the classical theory of elliptic and modular functions, nothing can surpass

A. Hurwitz, R. Courant, Vorlerungen über Allgemeine Funtionentheorie und Elliptische Funktionen, Part II, Springer-Verlag (1929).

For modular forms, a good introduction is:

B. Schoeneberg, Elliptic Modular Functions: An Introduction, Springer-Verlag (Grundlehren Band 203) (1974).

We have avoided, in this brief survey, the algebraic geometry of the objects being uniformised elliptic curves and the modular curves. Two general references are:

S. Lang, Elliptic Functions, Addison-Wesley (1973) where analytic and algebraic topics are mixed (but the series $\Sigma \exp(\pi i n^2 \tau)$ is scarcely mentioned), and

A. Robert, Elliptic Curves, Springer-Verlag Lecture Notes 326 (1973).

There are many open problems of very many kinds that could be mentioned. I want only to draw attention to several problems relating directly to theta functions, whose resolution would significantly clarify the theory.

(I) Which modular forms are polynomials in theta constants? More precisely:

Is every cusp form of wt. $n \geq 3$ a polynomial of degree $2n$ in the functions $\vartheta_{a,b}(0,\tau)$, $a,b \in \mathbb{Q}$?

(II) Can Jacobi's formula be generalized, e.g., to

$$(\frac{\partial}{\partial z}\vartheta_{a,b})(0,\tau) = \{\text{cubic polynomial in } \vartheta_{c,d}\text{'s}\}$$

for <u>all</u> $a,b \in \mathbb{Q}$? Similarly, are there generalizations of Jacobi's formula with higher order differential operators (see Ch. II, §7)?

(III) Can the modular forms $\vartheta_{a,b}(0,n\tau)$ be written, e.g., as

$$\frac{\text{Quadratic polyn. in } \vartheta_{c,d}\text{'s}}{\text{Linear polyn. in } \vartheta_{c,d}\text{'s}} ?$$

(IV) Can all relations among the $\vartheta_{a,b}(0,\tau)$'s be deduced from Riemann's theta relation, or generalizations thereof? A precise statement of this conjecture is given in Ch. II, §6.

# Chapter II: Basic results on theta functions in several variables

## §1. Definition of $\vartheta$ and its periodicity in $\vec{z}$.

We seek a generalization of the function $\vartheta(z,\tau)$ of Chapter I where $z \in \mathbb{C}$ is replaced by a g-tuple $\vec{z} = (z_1,\cdots,z_g) \in \mathbb{C}^g$, and which, like the old $\vartheta$, is quasi-periodic with respect to a lattice L but where $L \subset \mathbb{C}^g$. The higher-dimensional analog of $\tau$ is not so obvious. It consists in a symmetric g×g complex matrix $\Omega$ whose imaginary part is positive definite: why this is the correct generalization will appear later. Let $\mathfrak{H}_g$ be the set of such $\Omega$. Thus $\mathfrak{H}_g$ is an open subset in $\mathbb{C}^{g(g+1)/2}$. It is called the Siegel upper-half-space. The fundamental definition is:

$$\vartheta(\vec{z},\Omega) = \sum_{\vec{n}\in\mathbb{Z}^g} \exp(\pi i\,{}^t\vec{n}\Omega\vec{n} + 2\pi i\,{}^t\vec{n}\cdot\vec{z}).$$

(Here $\vec{n},\vec{z}$ are thought of as column vectors, so ${}^t\vec{n}$ is a row vector, ${}^t\vec{n}\cdot\vec{z}$ is the dot product, etc.; we shall drop the arrow where there is no reason for confusion between a scalar and a vector.)

Proposition 1.1. *$\vartheta$ converges absolutely and uniformly in $\vec{z}$ and $\Omega$ in each set*

$$\max_i |\text{Im } z_i| < \frac{c_1}{2\pi} \quad \text{and}$$

$$\text{Im } \Omega \geq c_2 I_g \ ;$$

*hence it defines a holomorphic function on* $\mathbb{C}^g \times \mathfrak{H}_g$.

Proof:

$$\left|\exp(\pi i\,{}^t\vec{n}\Omega\vec{n}+2\pi i\,{}^t\vec{n}\vec{z})\right| \le \exp(-\pi c_2(\Sigma n_i^2)+c_1\Sigma|n_i|) = \prod_{i=1}^{g}\exp(-\pi c_2 n_i^2+c_1|n_i|);$$

hence the series is dominated by $\left(\sum_{n\geq 0}\exp(-\pi c_2 n^2+c_1 n)\right)^g$ and

$\sum_{n\geq 0}\exp(-\pi c_2 n^2+c_1 n) = \text{const.} \sum_{n\geq 0}\exp\left[-\pi c_2\left(n-\frac{c_1}{2\pi c_2}\right)^2\right]$ which converges

like $\int_0^\infty e^{-x^2}dx$. Q.E.D.

Note that $\forall\Omega$, $\exists\vec{z}$ such that $\vartheta(\vec{z},\Omega) \neq 0$ because $\sum_{\vec{n}} e^{\pi i\,{}^t n\Omega n}e^{2\pi i\,{}^t nz}$ is a Fourier expansion of $\vartheta$, with Fourier coefficients $e^{\pi i\,{}^t n\Omega n} \neq 0$.

$\vartheta$ may be written more conceptually* as a series

$$\vartheta(\ell,Q) = \sum_{\vec{n}\in\mathbb{Z}^g}\exp(Q(\vec{n})+\ell(\vec{n}))$$

where $Q$ is a complex-valued quadratic function of $\vec{n}$ and $\ell$ is a complex-valued linear function of $\vec{n}$. To make this series converge, it is necessary and sufficient that Re $Q$ be positive definite. Then any such $Q$ is of the form

$$Q(\vec{x}) = \pi i\,{}^t x\cdot\Omega\cdot x, \quad \Omega\in\mathfrak{h}_g$$

and any such $\ell$ is of the form

$$\ell(\vec{x}) = 2\pi i\,{}^t x\cdot z, \quad z\in\mathbb{C}^g$$

* $\theta$ was explained this way in a lecture by Roy Smith.

hence any such $\vartheta(\ell, Q)$ equals $\vartheta(\vec{z}, \Omega)$. (This gives a formal justification for the introduction of $\mathfrak{h}_g$).

To $\Omega$, we now associate a lattice $L_\Omega \subset \mathbb{C}^g$:

$$L_\Omega = \mathbb{Z}^g + \Omega\mathbb{Z}^g$$

i.e., $L_\Omega$ is the lattice generated by the unit vectors and the columns of $\Omega$. The basic property of $\vartheta$ is to be "quasi-periodic" for $z \longmapsto z+a$, $a \in L_\Omega$. Here quasi-periodic means periodic up to a simple multiplicative factor. In fact,

$$\vartheta(\vec{z}+\vec{m}, \Omega) = \vartheta(\vec{z}, \Omega)$$

$$\vartheta(\vec{z}+\Omega\vec{m}, \Omega) = \exp(-\pi i\, {}^t m\Omega m - 2\pi i\, {}^t m z)\vartheta(z, \Omega) \qquad \forall \vec{m} \in \mathbb{Z}^g .$$

Proof. The 1st equality follows from the Fourier expansion of $\vartheta$ (with period 1); the 2nd one holds because of the symmetry of $\Omega$:

$$\exp[\pi i\,{}^t n\Omega n+2\pi i\,{}^t n(z+\Omega m)] = \exp(\pi i\,{}^t n\Omega n+\pi i\,{}^t n\Omega m+\pi i\,{}^t m\Omega n+2\pi i\,{}^t nz)$$

$$= \exp[\pi i\,{}^t(n+m)\Omega(n+m)+2\pi i\,{}^t(n+m)z-\pi i\,{}^t m\Omega m-2\pi i\,{}^t mz]$$

and, as $\vec{n}+\vec{m}$ varies over $\mathbb{Z}^g$, $\vec{n}$ does too; so

$$\sum_{n\in\mathbb{Z}^g} \exp[\pi i\,{}^t n\Omega n+2\pi i\,{}^t n(z+\Omega m)] = \exp(-\pi i\,{}^t m\Omega m-2\pi i\,{}^t mz).\sum_{n\in\mathbb{Z}^g} \exp(\pi i\,{}^t n\Omega n+2\pi i\,{}^t n z).$$

Q.E.D.

In fact, conversely, if $f(\vec{z})$ is an entire function such that

$$f(\vec{z}+\vec{m}) = f(\vec{z})$$

$$f(\vec{z}+\Omega\vec{m}) = \exp\left(-\pi i\,{}^t\vec{m}\Omega\vec{m}-2\pi i\,{}^t\vec{m}\,\vec{z}\right).f(\vec{z})$$

then $f(\vec{z}) = \text{const.}\,\vartheta(\vec{z},\Omega)$.

Proof: Because of the periodicities of $f(\vec{z})$ with respect to $\mathbb{Z}^g$, we can expand $f(\vec{z})$ in a Fourier series:

$$f(z) = \sum_{\vec{n}} c_{\vec{n}} \exp(2\pi i\,{}^t\vec{n}z);$$

now the second set of conditions gives us recursive relations among the coefficients $c_n$:

$$f(z+\Omega_k) = \sum_{n\in\mathbb{Z}^g} c_n \exp 2\pi i\, {}^t n(z+\Omega_k) = \sum c_n \exp(2\pi i\, {}^t n\Omega_k)\exp(2\pi i\, {}^t n\cdot z)$$

($\Omega_k = k^{\text{th}}$ column of $\Omega$), hence

$$\exp(-\pi i\Omega_{kk}-2\pi i z_k)\cdot \sum c_n \exp(2\pi i\, {}^t nz) = \sum c_n \exp(2\pi i\, {}^t n\Omega_k)\exp(2\pi i\, {}^t n\cdot z) .$$

Comparing the two Fourier expansions for $f(z+\Omega_k)$, we obtain:

$$c_{\vec{n}+\varepsilon_k} = c_{\vec{n}} e^{2\pi i\, {}^t\vec{n}\Omega_k+\pi i\Omega_{kk}}, \qquad \varepsilon_k = k^{\text{th}} \text{ unit vector.}$$

Thus $f$ is completely determined by the choice of the coefficient $c_0$.

Q.E.D.

This result suggests the following definition:

Definition 1.2. Fix $\Omega \in \mathfrak{H}_g$. Then an entire function $f(\vec{z})$ on $\mathbb{C}^g$ is $L_\Omega$-quasi-periodic of weight $\ell$ if

$$f(\vec{z}+\vec{m}) = f(\vec{z}),$$

$$f(\vec{z}+\Omega\cdot\vec{m}) = \exp(-\pi i\ell\cdot {}^t m\cdot\Omega\cdot m - 2\pi i\ell\cdot {}^t z\cdot m).f(\vec{z})$$

for all $\vec{m} \in \mathbb{Z}^g$. Let $R_\ell^\Omega$ be the vector space of such functions $f$.

As in the previous Chapter, one of the applications of such functions is to define holomorphic maps from the torus $\mathbb{C}^g/L_\Omega$ to projective space. In fact, if $f_0,\cdots,f_n$ are $L_\Omega$-quasi-periodic of the same weight $\ell$ and have the extra property that at every $\vec{a} \in \mathbb{C}^g$, $f_i(\vec{a}) \neq 0$ for at least one i, then

$$\vec{z} \longmapsto (f_0(\vec{z}), \cdots, f_n(\vec{z}))$$

defines a holomorphic map

$$\mathbb{C}^g/L_\Omega \longrightarrow \mathbb{P}^n .$$

By a slight generalization of $\vartheta$ known as the theta functions $\vartheta[{a \atop b}]$ with rational characteristics, we can easily find a basis of $R_\ell^\Omega$. These are just translates of $\vartheta$ multiplied by an elementary exponential factor:

$$\vartheta\begin{bmatrix}\vec{a}\\ \vec{b}\end{bmatrix}(\vec{z},\Omega) = \exp(\pi i\,{}^t\vec{a}\Omega\vec{a} + 2\pi i\,{}^t\vec{a}\cdot(\vec{z}+\vec{b}))\cdot\vartheta(\vec{z}+\Omega\vec{a}+\vec{b},\Omega)$$

$$\text{for all } \vec{a},\vec{b} \in \mathbb{Q}^g.$$

Written out, we have:

$$\vartheta\begin{bmatrix}\vec{a}\\ \vec{b}\end{bmatrix}(\vec{z},\Omega) = \sum_{\vec{n}\in\mathbb{Z}^g} \exp\left[\pi i\,{}^t(\vec{n}+\vec{a})\Omega(\vec{n}+\vec{a}) + 2\pi i\,{}^t(\vec{n}+\vec{a})(\vec{z}+\vec{b})\right] .$$

The original $\vartheta$ is just $\vartheta\begin{bmatrix}0\\0\end{bmatrix}$ and if $\vec{a},\vec{b}$ are increased by integral vectors, $\vartheta[{a \atop b}]$ hardly changes:

$$\vartheta\begin{bmatrix}\vec{a}+\vec{n}\\ \vec{b}+\vec{m}\end{bmatrix}(\vec{z},\Omega) = \exp(2\pi i\,{}^t\vec{a}\cdot\vec{m})\cdot\vartheta\begin{bmatrix}\vec{a}\\ \vec{b}\end{bmatrix}(\vec{z},\Omega) .$$

Finally, the quasi-periodicity of $\vartheta[{a \atop b}]$ is given by:

$$\vartheta\begin{bmatrix}\vec{a}\\ \vec{b}\end{bmatrix}(\vec{z}+\vec{m},\Omega) = \exp(2\pi i\,{}^t\vec{a}\cdot\vec{m})\cdot\vartheta\begin{bmatrix}\vec{a}\\ \vec{b}\end{bmatrix}(\vec{z},\Omega)$$

$$\vartheta\begin{bmatrix}\vec{a}\\ \vec{b}\end{bmatrix}(\vec{z}+\Omega\vec{m},\Omega) = \exp(-2\pi i\,{}^t\vec{b}\cdot\vec{m})\cdot\exp(-\pi i\,{}^t\vec{m}\Omega\vec{m} - 2\pi i\,{}^t\vec{m}\cdot\vec{z})\cdot\vartheta\begin{bmatrix}\vec{a}\\ \vec{b}\end{bmatrix}(\vec{z},\Omega)$$

which differs only by roots of unity from the law for $\vartheta\begin{bmatrix}0\\0\end{bmatrix}$. All of these identities are immediately verified by writing them out, but should be carefully checked and thought through on first acquaintance. Using these functions, we prove:

Proposition 1.3: Fix $\Omega \in \mathfrak{h}_g$. Then a basis of $R_\ell^\Omega$ is given by either:

i) $$f_{\vec{a}}(\vec{z}) = \vartheta\begin{bmatrix}\vec{a}/\ell\\0\end{bmatrix}(\ell\cdot\vec{z},\ell\cdot\Omega), \qquad 0 \le a_i < \ell$$

or

ii) $$g_{\vec{b}}(\vec{z}) = \vartheta\begin{bmatrix}0\\\vec{b}/\ell\end{bmatrix}(\vec{z}, \ell^{-1}.\Omega), \qquad 0 \le b_i < \ell \quad .$$

If $\ell = k^2$, then a 3rd basis is given by

iii) $$h_{\vec{a},\vec{b}}(\vec{z}) = \vartheta\begin{bmatrix}\vec{a}/k\\\vec{b}/k\end{bmatrix}(\ell.\vec{z},\Omega), \qquad 0 \le a_i,b_i < k.$$

These bases are related by

$$g_{\vec{b}} = \sum_{\vec{a}} \exp(2\pi i\, \ell^{-1}.\,{}^t\vec{a}\cdot\vec{b})\cdot f_{\vec{a}}$$

$$h_{\vec{a},\vec{b}} = \sum_{\vec{c}\equiv\vec{a} \bmod k} \exp(2\pi i k^{-1}.\,{}^t\vec{c}\cdot\vec{b}) f_{\vec{c}} \ .$$

Proof: As above, we expand functions in $R^\Omega_\ell$ as Fourier series in $\vec{z}$. By quasi-periodicity with respect to $\Omega\cdot\mathbb{Z}^g$, we check that a function f lies in $R^\Omega_\ell$ if and only if f can be expressed as

$$f_\chi(\vec{z}) = \sum_{\vec{n}\in\mathbb{Z}^g} \chi(\vec{n})\exp(\pi i\ell^{-1}\cdot{}^t\vec{n}\cdot\Omega\cdot\vec{n} + 2\pi i\,{}^t\vec{n}\cdot\vec{z})$$

where $\chi$ is constant on cosets of $\ell\cdot\mathbb{Z}^g$. Taking $\chi$ to be the characteristic function of $\vec{a}+k\mathbb{Z}^g$, $f_\chi$ becomes $f_{\vec{a}}$; taking $\chi$ to be the character $\vec{n} \longmapsto \exp(2\pi i\ell^{-1}\cdot{}^t\vec{n}\cdot\vec{b})$, $f_\chi$ becomes $g_{\vec{b}}$; and if $\ell = k^2$, taking $\chi$ to be the restriction of $\vec{n} \longmapsto \exp(2\pi i\ell^{-1}\cdot{}^t\vec{n}\cdot\vec{b})$ to $\vec{a}+k\cdot\mathbb{Z}^g$, $f_\chi$ becomes $h_{\vec{a},\vec{b}}$. QED

Let us see how these functions can be used projectively to embed not only $\mathbb{C}^g/L_\Omega$ but "isogenous" tori $\mathbb{C}^g/L$, $L$ a lattice in $L_\Omega\cdot\mathbb{Q}$.

First some notation: fix $\Omega \in \mathfrak{H}_g$ and identify $\mathbb{R}^g \times \mathbb{R}^g$ with $\mathbb{C}^g$ via $\Omega$ by

$$\alpha_\Omega\colon \mathbb{R}^g \times \mathbb{R}^g \longrightarrow \mathbb{C}^g,\ (\vec{x},\vec{y}) \longmapsto \Omega\vec{x}+\vec{y} = \vec{z}.$$

Note then that $\alpha_\Omega$ identifies $\mathbb{Z}^g\times\mathbb{Z}^g$ with $L_\Omega = \mathbb{Z}^g + \Omega\mathbb{Z}^g$. Define

$$e\colon \mathbb{R}^{2g} \times \mathbb{R}^{2g} \longrightarrow \mathbb{C}_1^*,\quad e(\vec{x},\vec{y}) = \exp 2\pi i A(\vec{x},\vec{y})$$

where $A$ is the real skew-symmetric form on $\mathbb{R}^{2g}\times\mathbb{R}^{2g}$ defined by

$$A(\vec{x},\vec{y}) = {}^t\vec{x}_1\cdot\vec{y}_2 - {}^t\vec{y}_1\cdot\vec{x}_2,\quad \vec{x} = (\vec{x}_1,\vec{x}_2),\quad \vec{y} = (\vec{y}_1,\vec{y}_2).$$

It is immediate that

$$e(x,x) = 1, \; e(x,y) = e(y,x)^{-1}$$

and

$$e(x+x',y) = e(x,y)e(x',y).$$

Thus e is bi-multiplicative and so we can talk of the <u>perpendicular</u> $V^{\perp}$ of a subset $V \subseteq \mathbb{R}^{2g}$, namely,

$$V^{\perp} = \{x \in \mathbb{R}^{2g} \mid e(x,a) = 1, \forall a \in V\}.$$

We shall be particularly interested in the perpendiculars $L^{\perp}$ within $\mathbb{Q}^{2g}$ of lattices $L \subset \mathbb{Q}^{2g}$, i.e.,

$$L^{\perp} = \{x \in \mathbb{Q}^{2g} \mid e(x,a) = 1, \forall a \in L\}.$$

It is immediate that

(i) $\quad (\mathbb{Z}^{2g})^{\perp} = \mathbb{Z}^{2g}$, $\quad$ (ii) $(\frac{1}{n}\mathbb{Z}^{2g})^{\perp} = n\mathbb{Z}^{2g}$, $n \in \mathbb{N}$.

In fact, more generally, for lattices $L, L_1, L_2$ in $\mathbb{Q}^{2g}$, we have:

$$(\tfrac{1}{n}L)^{\perp} = nL^{\perp}, \; (L^{\perp})^{\perp} = L, \; L_1 \subseteq L_2 \iff L_1^{\perp} \supseteq L_2^{\perp}, \text{ etc.}$$

In particular, if $L \subseteq \mathbb{Z}^{2g}$ is of index s, then $\mathbb{Z}^{2g} \subseteq L^{\perp}$ is of index s. Further, notice that in this case $\alpha_{\Omega}(L) \subseteq L_{\Omega} = \alpha_{\Omega}(\mathbb{Z}^{2g})$ is also of index s in $L_{\Omega}$. Let $a_i, b_i \in L^{\perp}$, $1 \leq i \leq s$, be coset representatives of $L^{\perp}/\mathbb{Z}^{2g}$. Let us call the set

$$B_\Omega(L) = \left\{ z \in \mathbb{C}^g \,\middle|\, \vartheta\begin{bmatrix} a_i \\ b_i \end{bmatrix}(z,\Omega) = 0,\ 1 \le i \le s \right\} \Big/ \alpha_\Omega(L)$$

the set of "base points" in the complex torus $\mathbb{C}^g/\alpha_\Omega(L)$. Now the "rational morphism" $\varphi_L$ is given in the

Proposition 1.2. For all $L \subset \mathbb{Z}^{2g}$ of index $s$, via the $\vartheta\begin{bmatrix} a_i \\ b_i \end{bmatrix}$'s, we have a canonically defined holomorphic map

$$\varphi_L:\ [\mathbb{C}^g/\alpha_\Omega(L)] - B_\Omega(L) \longrightarrow \mathbb{P}^{s-1}$$

namely

$$\varphi_L(z) = (\ldots, \vartheta\begin{bmatrix} a_i \\ b_i \end{bmatrix}(z,\Omega), \ldots).$$

We have only to check that $\varphi_L$ is well-defined. We do this as follows: let $(a,b) \in L^\perp$ and $(a',b') \in L$. Then by the quasi-periodicity of $\vartheta\begin{bmatrix} a \\ b \end{bmatrix}$:

$$\vartheta\begin{bmatrix} a \\ b \end{bmatrix}(z+\alpha_\Omega(a',b'),\Omega) = \vartheta\begin{bmatrix} a \\ b \end{bmatrix}(z+\Omega a'+b',\Omega)$$

$$= \exp[2\pi i\,{}^t a\cdot b' - 2\pi i\,{}^t b\cdot a' - \pi i\,{}^t a'\Omega a' - 2\pi i\,{}^t a' z]\,\vartheta\begin{bmatrix} a \\ b \end{bmatrix}(z,\Omega)$$

$$= \lambda(a',b',z).e((a,b),(a',b')).\,\vartheta\begin{bmatrix} a \\ b \end{bmatrix}(z,\Omega).$$

But $e((a,b),(a',b')) = 1$ and $\lambda(a',b',z)$ is independent of $a,b$ and this proves the result.

The main result concerning this map is this:

Theorem 1.3 (Lefschetz): Let $L \subseteq \mathbb{Z}^{2g}$ be a lattice of index s and assume that $L \subseteq r.L^{\perp}$ for some $r \in \mathbb{N}$. Then:

(1) if $r \geq 2$, $B_{\Omega}(L) = \emptyset$, i.e., $\varphi_L$ is defined on all of $\mathbb{C}^g/\alpha_{\Omega}(L)$,

(2) if $r \geq 3$, $\varphi_L$ is an embedding and the image is an algebraic subvariety of $\mathbb{P}^{s-1}$, i.e., the complex torus $\mathbb{C}^g/\alpha_{\Omega}(L)$ is embedded as an algebraic subvariety of $\mathbb{P}^{s-1}$;

(3) every complex torus that can be embedded in a projective space (or, more generally, whose points can be separated by meromorphic functions) is isomorphic to $\mathbb{C}^g/\alpha_{\Omega}(L)$ for some $\Omega \in \mathfrak{H}_g$ and some L.

For a full proof, the reader may consult §3 of the author's book, Abelian Varieties, Oxford University Press (1974). Here we shall skip completely the proof of (3) and outline the proofs of (1) and (2). Note that it is (3) which explains why we have focussed attention on the special type of lattice $L_{\Omega}$ in $\mathbb{C}^g$: these and their sublattices are the only ones which lead to complex tori which are also algebraic varieties. It would be impossible to find entire functions f(z) quasi-periodic for more general lattices because of this result.

The first step in the proof of (1) and (2) is:

Lemma 1.4: Let f(z) be any holomorphic function such that

$$f(z+\Omega a'+b') = \exp(-\pi i\,{}^t a'\Omega a' - 2\pi i\,{}^t a' z)\cdot f(z)$$

for all $(a',b') \in L$. Then f(z) is a linear combination of the functions $\vartheta\begin{bmatrix} a_i \\ b_i \end{bmatrix}(z,\Omega)$, $1 \leq i \leq s$.

This is proven by a variation of the argument used to prove that $\vartheta$ is characterized up to a scalar by its functional equation: one makes a Fourier expansion of f for the lattice $\alpha_\Omega(L) \cap \mathbb{Z}^g$ in $\mathbb{C}^g$, and expresses the remaining functional equations as recursion relations on the Fourier coefficients. These leave only s coefficients to be determined and it's easy to see that the s functions $z \longmapsto \vartheta\begin{bmatrix}a_i\\b_i\end{bmatrix}(z,\Omega)$ are linearly independent. The $2^{nd}$ step is a little symplectic geometry over $\mathbb{Z}$:

<u>Lemma 1.5</u>: <u>For all</u> $L \subset \mathbb{Z}^{2g}$ <u>such that</u> $L \subset rL^\perp$, <u>there is a lattice</u> $L_1$ <u>with</u> $L \subset L_1$ <u>and</u> $L_1 = rL_1^\perp$. <u>Such an</u> $L_1$ <u>has a standard basis</u>:

$$(o,e_1),\cdots,(o,e_g),(f_1',f_1''),\cdots,(f_g',f_g'')$$

<u>with</u>

$$A((o,e_i),(f_j',f_j'')) = -r\cdot\delta_{ij}$$

$$A((f_i',f_i''),(f_j',f_j'')) = 0.$$

<u>In fact, we can even find 2 such enlargements</u>: $L \subset L_1$, $L \subset L_1'$ <u>such that</u> $L = L_1 \cap L_1'$ <u>and</u> $L_1 = rL_1^\perp$, $L_1' = rL_1'^\perp$.

Thus $e_1,\cdots,e_g$ and $\underline{f}_1 = \Omega f_1'+f_1'',\cdots,\underline{f}_g = \Omega f_g'+f_g''$ are a basis of $\alpha_\Omega(L_1)$. We then define a linear map

$$S:\ \mathbb{C}^g \longrightarrow \mathbb{C}^g$$

by requiring that:

$$S(e_i) = (0,\cdots,1,\cdots,0), \text{ the } i^{th} \text{ unit vector.}$$

Let

$$S(\underline{f}_i) = (\Omega_{1i}',\cdots,\Omega_{gi}').$$

In matrix notation, if we write $E, F', F''$ for the matrices whose columns are $e_i, f_i'$ and $f_i''$, then the lemma says

$$^tE.F' = rI$$

$$^tF'.F'' = {}^tF''.F' \quad \text{or} \quad {}^tF'.F'' \text{ symmetric.}$$

Thus $S$ is given by the matrix $E^{-1}$ and

$$\begin{aligned}\Omega' &= SF \\ &= S(\Omega F' + F'') \\ &= \frac{1}{r}({}^tF'\Omega F' + {}^tF'\cdot F'')\end{aligned}$$

so $\Omega'$ is again symmetric with positive definite imaginary part. Note that the linear map $S$ carries the lattice $\alpha_\Omega(L_1)$ to the lattice $L_{\Omega'}$. The purpose of this construction is to take the function $\vartheta(z,\Omega')$ quasi-periodic with respect to the lattice $L_{\Omega'}$ and form from it the function $\vartheta(Sz,\Omega')$, quasi-periodic for $\alpha_\Omega(L_1)$. We check:

<u>Lemma 1.6:</u> <u>For suitable $\vec{b} \in \mathbb{C}^g$, the function</u>

$$f(z) = \vartheta(Sz+b,\Omega')$$

<u>satisfies:</u>

a) $f(z+e_i) = f(z)$,

b) $f(z+\Omega f_i' + f_i'') = \exp(-\frac{\pi i}{r}\, {}^tf_i'\Omega f_i' - \frac{2\pi i}{r}\, {}^tf_i' z)\cdot f(z)$

<u>for</u> $1 \leq i \leq g$. <u>Therefore for all</u> $\alpha_1,\cdots,\alpha_r \in \mathbb{C}^g$ <u>such that</u> $\sum_{i=1}^r \alpha_i = 0$, <u>the functions</u> $g(z) = \prod_{i=1}^r f(z+\alpha_i)$ <u>satisfy</u>

a) $g(z+e_i) = g(z)$

b) $g(z+\Omega f_i'+f_i'') = \exp(-\pi i\, {}^t f_i'\Omega f_i' - 2\pi i\, {}^t f_i' z)\, g(z)$,

*for* $1 \le i \le g$, *hence* $g$ *satisfies the hypothesis of lemma 1.4*.

The proof is quite straightforward. Without any $\vec{b}$, we use the functional equation for $\vartheta$ to find the law for f, but come up with a root of unity in (b). Adjusting the $\vec{b}$, we get rid of these roots of unity. The second part is an immediate consequence of the first. This is now Lefschetz's central idea: that there is a related $\vartheta$-function f such that *all* products

$$f(z+\alpha_1)\cdot\cdots\cdots\cdot f(z+\alpha_r)$$

(for $\alpha_1+\cdots+\alpha_r = 0$) are linear combinations of the functions $\vartheta\begin{bmatrix}a_i\\b_i\end{bmatrix}(z,\Omega)$. We next show that:

*Lemma 1.7:* a) *If* $r \ge 2$, *then for all* $u \in \mathbb{C}^g$, *there is a product* $g(z)$ *as above such that* $g(u) \ne 0$.

b) *If* $r \ge 3$, *then for all* $u,v \in \mathbb{C}^g$ *with* $u-v \notin \alpha_\Omega(L_1)$, *there is a linear combination* $h(z) = \sum c_i g_i(z)$ *of products* $g_i(z)$ *as above such that* $h(u) = 0$, $h(v) \ne 0$.

c) *If* $r \ge 3$, *then for all* $u \in \mathbb{C}^g$, *and tangent vector* $\sum d_i \frac{\partial}{\partial z_i} \ne 0$, *there is a linear combination* $h(z) = \sum c_i g_i(z)$ *of products* $g_i(z)$ *as above such that* $h(u) = 0$, $\sum d_i \frac{\partial h}{\partial z_i}(u) \ne 0$.

This clearly finishes the proof: by (a), $B_\Omega(L) = \emptyset$ if $r \geq 2$. By (b), if $x,y \in \mathbb{C}^g/\alpha_\Omega(L)$ and $\varphi_L(x) = \varphi_L(y)$, then provided $r \geq 3$, $x-y \in \alpha_\Omega(L_1/L)$. Applying the same argument to $L_1'$ and products $g'$ constructed similarly, we deduce $x-y \in \alpha_\Omega(L_1'/L)$ too. Thus $x = y$. By (c), the differential of $\varphi_L$ is one-one if $r \geq 3$ too. The proof of the lemma is not difficult: we take $r = 3$ for simplicity of notation and see how (b) is proven. For the other parts, we refer the reader to [AV, pp. 30-33]. To prove (b), take $u,v \in \mathbb{C}^g$ and assume $h(u) = 0 \Longrightarrow h(v) = 0$. Then there is a complex number $\gamma$ such that for all $a,b \in \mathbb{C}^g$,

$$(*) \qquad f(v+a)f(v+b)f(v-a-b) = \gamma f(u+a)f(u+b)f(u-a-b).$$

This is because the linear functionals which carry the function $f(z+a)f(z+b)f(z-a-b)$ to its values at $u$ and at $v$ must be multiples of each other if one is zero whenever the other is zero. Now in (*), take logs and differentiate with respect to $a$. If $\omega$ is the meromorphic 1-form $df/f$, we find

$$\omega(v+a) - \omega(v-a-b) = \omega(u+a) - \omega(u-a-b), \text{ all } a,b \in \mathbb{C}^g.$$

Thus $\omega(v+z)-\omega(u+z)$ is independent of $z$, hence is a constant 1-form $2\pi i \sum c_i dz_i$. But $\omega(v+z)-\omega(u+z) = d \log f(v+z)/f(u+z)$, so this means that

$$f(z+v-u) = c_0 e^{2\pi i {}^t\vec{c}\cdot\vec{z}} f(z)$$

for some constant $c_0$. In this formula, you substitute $z+e_i$ and $z+\Omega f_i'+f_i''$ for $z$ and use Lemma 1.6. It follows that

$$^t c.e_i \in \mathbb{Z}$$

$$\frac{^t f_i'.(u-v)}{r} \equiv {}^t c\cdot(\Omega f_i' + f_i'') \bmod \mathbb{Z}\,.$$

Now write $u-v = \Omega\mathbf{x}+y$, $x,y \in \mathbb{R}^g$. Take imaginary parts in the 2nd formula, to get

$$\frac{^t f_i'.\mathrm{Im}\,\Omega.x}{r} = {}^t c\cdot \mathrm{Im}\,\Omega\cdot f_i', \quad \text{all } i.$$

Hence $c = x/r$. Putting this back, we find:

$$^t(\tfrac{x}{r})\cdot e_i \in \mathbb{Z}$$

$$^t f_i'\cdot(\tfrac{y}{r}) - {}^t f_i''\cdot(\tfrac{x}{r}) \in \mathbb{Z}\,.$$

This means that $(\frac{x}{r},\frac{y}{r}) \in L_1^{\perp}$, or $(x,y) \in rL_1^{\perp} = L_1$, hence $u-v = \alpha_\Omega(x,y) \in \alpha_\Omega(L_1)$. This proves (b).

For further details, we refer the reader to [AV,§3]. Finally, what can we say about the complex tori $\mathbb{C}^g/\alpha_\Omega(L)$ for $L \subset \mathbb{Z}^{2g}$ such that $L \not\subset rL^{\perp}$, $r \geq 2$, or even for arbitrary lattices $L \subset \mathbb{Q}^{2g}$? In fact, we do not get more general complex tori in this way, because of the isomorphism:

$$\mathbb{C}^g/\alpha_\Omega(L) \xrightarrow{\approx} \mathbb{C}^g/\alpha_\Omega(nL)$$

given by

$$\vec{z} \longmapsto n\vec{z}.$$

Because of these isomorphisms, the theorem has the Corollary:

Corollary: A complex torus $\mathbb{C}^g/L$ can be embedded in projective space if and only if

$$A(L) \subset \Omega\mathbb{Q}^g + \mathbb{Q}^g$$

for some $g \times g$ complex matrix A, and some $\Omega \in \mathfrak{H}_g$.

## §2. The Jacobian Variety of a Compact Riemann Surface.

It is hard on first sight to imagine how the higher-dimensional generalizations of $\vartheta$ of the last section were discovered. The main result showed that the complex tori $\mathbb{C}^g/L_\Omega$ and their finite coverings $\mathbb{C}^g/L$, ($L \subset L_\Omega$ of finite index) could be embedded by theta functions in projective space, but no other tori can be so embedded. This justifies after the fact considering only the lattices $L_\Omega$ built up via $\Omega \in \mathfrak{H}_g$. If $g \geq 2$, these are quite special: $\Omega$ has $\frac{g(g+1)}{2}$ complex parameters, whereas a general lattice is of the form

$$\mathbb{Z}^g + \Omega\cdot\mathbb{Z}^g$$

$\Omega$ any $g\times g$ complex matrix with $\det(\mathrm{Im}\,\Omega) \neq 0$, hence it depends on $g^2$ complex parameters. However, these particular tori arose in the 19th century from a very natural source: as Jacobian Varieties of compact Riemann Surfaces. Much of the theory of theta functions is specifically concerned with the identities that arise from this set-up. The point of this section is to explain briefly the beginnings of this theory. In Chapter III, we will study it much more extensively in the very particular case of hyperelliptic Riemann Surfaces.

Let X be a compact Riemann Surface. As a topological space, X is a compact orientable 2-manifold , hence it is determined, up to diffeomorphism, by its genus g, i.e., the number of "handles". It is well known that the genus g occurs in at least 3 other fundamental roles in the description of X. We shall assume the basic existence theorem: that g is also the dimension of the vector space of holomorphic 1-forms on $X^*$. An extensive treatment of this, as well as the other topics in this section, can be found in Griffiths-Harris, *Principles of Algebraic Geometry*.

The first step is to analyze the periods of the holomorphic 1-forms, by use of Green's theorem. To do this we have to dissect the 2-manifold X in some standard way: i.e., we want to cut X open on 2g disjoint simple closed paths, all beginning and ending at the same base point, so that what remains is a 2-cell. Then conversely, X can be reconstructed by starting with a polygon with 4g sides (one side each for the left and right sides of each path) and glueing these together in pairs, in particular all vertices being glued. The standard picture is this, drawn with $g = 3$:

*These are the differential forms $\omega$ locally given in analytic coordinates by $\omega = a(z)dz$, $a(z)$ holomorphic.

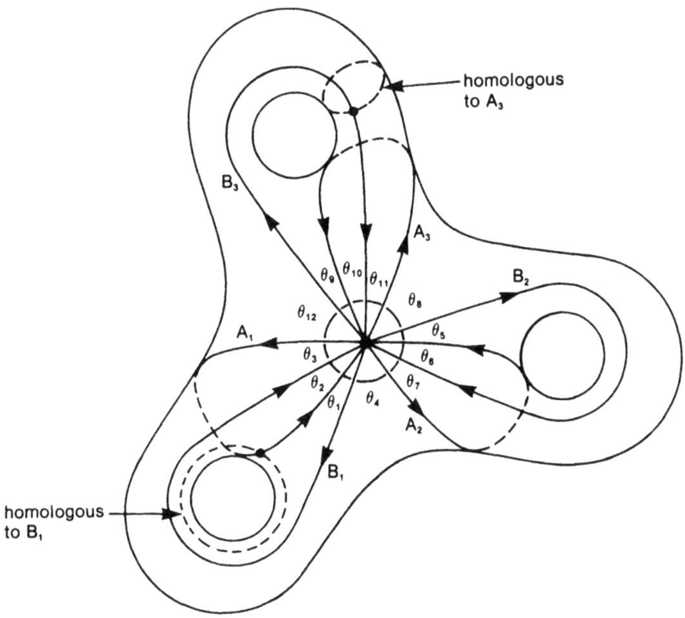

homologous
to A3
B3
A3
θ9
θ10
θ11
B2
θ8
θ12
A1
θ5
θ3
θ6
θ2
θ7
θ1
θ4
A2
B1
homologous
to B1

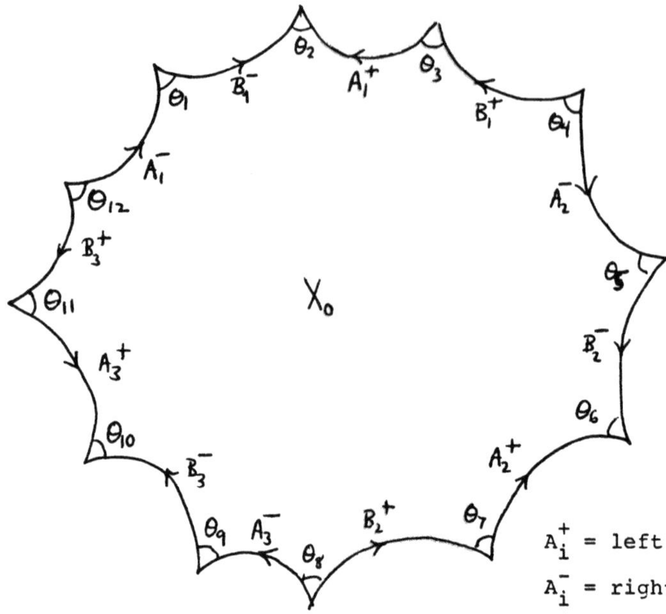


$A_i^+$ = left side of $A_i$

$A_i^-$ = right side of $A_i$

$B_i^+$ = left side of $B_i$

$B_i^-$ = right side of $B_i$

$\partial X_o = -\sum A_i^+ - \sum B_i^+ + \sum A_i^- + \sum B_i^-$

$$X_0 = X - \bigcup_{i=1}^{g} A_i - \bigcup_{i=1}^{g} B_i$$

Note that if $I(\sigma,\tau)$ is the intersection product of 2 cycles $\sigma,\tau$, then

$$I(A_i,A_j) = I(B_i,B_j) = 0$$

$$I(A_i,B_j) = \delta_{ij}.$$

Theorem 2.1. (Bilinear relations of Riemann): *Let* X *be a compact Riemann surface of genus* g, *with canonical dissection*

$$X = X_0 \amalg A_1 \amalg \cdots \amalg A_g \amalg B_1 \amalg \cdots \amalg B_g$$

*as above.*

a) *for all holomorphic* 1-*forms* $\omega,\eta$,

$$\sum_{i=1}^{g} \int_{A_i}\omega \cdot \int_{B_i}\eta - \sum_{i=1}^{g} \int_{B_i}\omega \cdot \int_{A_i}\eta = 0$$

b) *for all holomorphic* 1-*forms* $\omega$,

$$\operatorname{Im}\left(\sum_{i=1}^{g} \overline{\int_{A_i}\omega} \cdot \int_{B_i}\omega\right) > 0$$

Note that if we let $\Gamma(X,\Omega^1)$ denote the vector space of holomorphic 1-forms, and if we define the period map

$$\text{per}: \Gamma(X,\Omega^1) \longrightarrow \operatorname{Hom}(H_1(X,\mathbb{Z}),\mathbb{C}) = H^1(X,\mathbb{C})$$

by

$$\omega \longmapsto \left\{\text{the co-cycle } \sigma \longmapsto \int_\sigma \omega\right\}$$

then (a) can be interpreted as saying that with respect to cup product, the image of per is an isotropic subspace of $H^1(X,\mathbb{C})$.

In fact, $\cup$ on $H^1$ is dual to I on $H_1$. Thus if we associate to $\omega$ the 1-cycle

$$d(\omega) = \sum_{i=1}^{g}\left(\int_{B_i}\omega\right)\cdot A_i - \sum_{i=1}^{g}\left(\int_{A_i}\omega\right) B_i$$

then $d(\omega)$ satisfies

$$I(d(\omega),c) = \int_c \omega, \text{ all 1-cycles } C.$$

So $\text{per}(\omega) \cup \text{per}(\eta)$ is equal to $I(d(\omega),d(\eta))$, which is exactly the thing which (a) says is zero.

To prove the theorem, since $X_0$ is simply connected, there is a holomorphic function f on $X_0$ such that $\omega = df$. Then $f\cdot\eta$ is a closed 1-form, so by Green's theorem

$$\begin{aligned}
0 &= \int_{X_0} d(f\eta) \\
&= \int_{\partial X_0} f\eta \\
&= \sum_i \left(-\int_{A_i^+} f\eta + \int_{A_i^-} f\eta - \int_{B_i^+} f\eta + \int_{B_i^-} f\eta\right) \\
&= \sum_i \int_{A_i} [-(f \text{ on } A_i^+)+(f \text{ on } A_i^-)]\eta \\
&\quad + \sum_i \int_{B_i} [-(f \text{ on } B_i^+)+(f \text{ on } B_i^-)]\eta .
\end{aligned}$$

As df has no discontinuity on $A_i$ or $B_i$, f on $A_i^+$ must differ from f on $A_i^-$ by a constant, and likewise for $B_i^+, B_i^-$. But the path $B_i$ leads from $A_i^-$ to $A_i^+$ (see dashed line in diagram above) and the path $A_i$ leads from $B_i^+$ to $B_i^-$. Thus

$$0 = \sum_i \int_{A_i} \left( - \int_{B_i} \omega \right) \eta + \sum_i \int_{B_i} \left( + \int_{A_i} \omega \right) \eta$$

which proves (a). As for (b)

$$\int_{X_0} d(\bar{f}\omega) = \int_{\partial X_0} \bar{f}\omega = - \sum_i \overline{\int_{B_i} \omega} \cdot \int_{A_i} \omega + \sum_i \overline{\int_{A_i} \omega} \cdot \int_{B_i} \omega$$

as before . The right-hand side is

$$2i \ \mathrm{Im} \left( \sum_i \overline{\int_{A_i} \omega} \cdot \int_{B_i} \omega \right)$$

and $d(\bar{f}\omega) = d\bar{f} \wedge df$. Whenever f is a local analytic coordinate, let $f = x+iy$, x,y real coordinates. Then

$$d\bar{f} \wedge df = (dx-idy) \wedge (dx+idy)$$

$$= 2i \ dx \wedge dy.$$

Since $dx \wedge dy$ is a positive 2-form in the canonical orientation

$$\mathrm{Im} \int_X d\bar{f} \wedge df > 0.$$

QED

If we now introduce a canonical basis in $\Gamma(X,\Omega^1)$, a matrix $\Omega$ in Siegel's upper half space appears immediately:

Corollary 2.2. We can find a normalized basis $\omega_i$ of $\Gamma(X,\Omega^1)$ such that

$$\int_{A_i} \omega_j = \delta_{ij} .$$

Let $\Omega_{ij} = \int_{B_i} \omega_j$. Then $\Omega_{ij} = \Omega_{ji}$ and Im $\Omega_{ij}$ is positive definite.

Proof: The pairing between $\omega$'s of $1^{st}$ kind and $A_i$'s is non-degenerate because of b) in 2.1. By applying a) to $\omega = \omega_j$, $\eta = \omega_i$ we get $\Omega_{ji} - \Omega_{ij} = 0$; finally, in order to prove that, for any $\alpha_1,\ldots,\alpha_g$ real, $\mathrm{Im} \sum_{i,k} \alpha_i \Omega_{ik} \alpha_k > 0$, we let $\omega = \sum \alpha_i \omega_i$. By b)

$0 < \mathrm{Im} \sum_i \alpha_i ( \sum_k \alpha_k \Omega_{ki})$. QED

We may understand the situation in another way if we view the periods of 1-forms as a map:

$$\begin{aligned} \text{per}' : H_1(X,\mathbb{Z}) &\longrightarrow \mathrm{Hom}(\Gamma(X,\Omega^1),\mathbb{C}) \\ \sigma &\longmapsto \left\{ \text{the linear map } \omega \longmapsto \int_\sigma \omega \right\} \end{aligned}$$

or, if we use the basis $\omega_1,\cdots,\omega_g$ of $\Gamma(X,\Omega^1)$:

$$\begin{aligned} \text{per}' : H_1(X,\mathbb{Z}) &\longrightarrow \mathbb{C}^g \\ \sigma &\longmapsto \left( \int_\sigma \omega_1, \cdots, \int_\sigma \omega_g \right) . \end{aligned}$$

Corollary 2.3. *The map* per': $H_1(X,\mathbb{Z}) \longrightarrow \mathbb{C}^g$ *is injective and its image is the lattice* $L_\Omega$ *generated by integral vectors and the columns of* $\Omega$.

The fundamental construction of the classical theory of compact Riemann Surfaces is the introduction of the complex torus:

$$\mathrm{Jac}(X) \underset{\overline{\overline{\mathrm{def}}}}{} \mathbb{C}^g/L_\Omega .$$

By Corollary 2.3, if $P_o$ is a base point on X, then we obtain a holomorphic map

$$\begin{aligned} X &\longrightarrow \mathrm{Jac}(X) \\ P &\longmapsto \left(\int_{P_o}^{P} \omega_1, \cdots, \int_{P_o}^{P} \omega_g\right) \text{ mod periods.} \end{aligned}$$

This is well-defined: pick any path $\gamma$ from $P_o$ to P and evaluate all the integrals along $\gamma$. If $\gamma$ is changed, the vector of integrals is altered by a period, i.e., a vector in $L_\Omega$. More generally, if

$$\mathfrak{A} = \sum k_i P_i$$

is a cycle of points on X of degree 0, i.e., $\sum k_i = 0$, then we can associate to $\mathfrak{A}$ a point

$$I(\mathfrak{A}) \in \mathrm{Jac}(X)$$

given by

$$I(\mathfrak{A}) = \left(\int_\sigma \omega_1, \cdots, \int_\sigma \omega_g\right) \text{ mod periods,}$$

$\sigma$ a 1-chain on X so that $\partial\sigma = \sum k_i P_i$. The map $\mathfrak{A} \longmapsto I(\mathfrak{A})$ plays a central role in the function-theory on X because of the simple observation:

<u>Proposition</u> 2.4: <u>If</u> f <u>is a meromorphic function on</u> X <u>with poles</u> $\sum_{i=1}^{d} P_i$ <u>and zeroes</u> $\sum_{i=1}^{d} Q_i$ <u>(counted with multiplicities)</u>, <u>then</u>

$$I\left(\sum_{i=1}^{d} P_i - \sum_{i=1}^{d} Q_i\right) = 0.$$

<u>Proof</u>: For all $t \in \mathbb{C}$, let $D(t)$ be the cycle of points where f takes the value t, i.e., the fibre of the holomorphic map of degree d

$$f\colon X \longrightarrow \mathbb{P}^1$$

over t. If $P_o$ is a base point, consider $I(D(t)-d.P_o)$ as a function of t. Because the endpoints are varying analytically, so does

$$\int_{dP_o}^{D(t)} \omega_i$$

hence $t \longmapsto I(D(t)-dP_o)$ is a holomorphic map

$$\delta\colon \mathbb{P}^1 \longrightarrow \mathrm{Jac}(X).$$

But $\mathbb{P}^1$ is simply connected, so this map lifts to

$$\tilde{\delta}\colon \mathbb{P}^1 \longrightarrow \mathbb{C}^g.$$

Since there are no meromorphic functions on $\mathbb{P}^1$ without poles, except constants, $\tilde{\delta}$ and $\delta$ are constant. In particular $\delta(0) = \delta(\infty)$ and $\delta(\infty)-\delta(0) = I(\sum P_i - \sum Q_i)$. <u>QED</u>

The beautiful result which is the cornerstone of this theory is:

Theorem of Abel: *Given cycles* $\sum_{i=1}^{d} P_i$, $\sum_{i=1}^{d} Q_i$ *of the same degree, then conversely if* $I(\sum P_i - \sum Q_i) = 0$, *there is a meromorphic function* f *on* X *with poles* $\sum P_i$, *zeroes* $\sum Q_i$.

We shall prove this in the next section.

## §3. $\vartheta$ and the function theory on a compact Riemann Surface.

We continue to study a compact Riemann Surface X. As before, we fix a basis $\{A_i, B_i\}$ of $H_1(X,\mathbb{Z})$, obtaining a dual basis $\omega_i$ of holomorphic 1-forms, a period matrix $\Omega \in \mathfrak{H}_g$, and the Jacobian $\mathrm{Jac}(X) = \mathbb{C}^g/L_\Omega$. We also fix a base point $P_0 \in X$. By the methods of §1, we have the function $\vartheta(\vec{z},\Omega)$ on $\mathbb{C}^g$, quasi-periodic with respect to $L_\Omega$. We now ask:

1) Starting with $\vartheta(\vec{z},\Omega)$, what meromorphic functions on Jac(X) can we form?

2) Via the canonical map

$$\begin{array}{ccc} X & \longrightarrow & \mathrm{Jac}(X) \\ P & \longmapsto & \displaystyle\int_{P_0}^{P} \vec{\omega} \end{array}$$

what meromorphic functions on X can we form?

Starting with (1), we may allow $\Omega$ to be an arbitrary period matrix in $\mathfrak{H}_g$. Then there are 3 quite different ways in which we can form $L_\Omega$-periodic meromorphic functions on $\mathbb{C}^g$, from the $L_\Omega$-quasi-periodic but holomorphic function $\vartheta$.

Method I:

$$f(\vec{z}) = \frac{\prod_{i=1}^{n} \vartheta(\vec{z}+\vec{a}_i,\Omega)}{\prod_{i=1}^{n} \vartheta(\vec{z}+\vec{b}_i,\Omega)},$$

where $\vec{a}_i, \vec{b}_i \in \mathbb{C}^g$ are such that $\sum \vec{a}_i \equiv \sum \vec{b}_i \bmod \mathbb{Z}^g$, is a meromorphic function on $X_\Omega$, since the denominator doesn't vanish.

identically and the condition $\sum \vec{a}_i \equiv \sum \vec{b}_i$ provides us with $\Omega$-invariance:

$$f(\vec{z}+\Omega\vec{m}) = \frac{\exp(-\sum_i [\pi i\, {}^t\vec{m}\Omega\vec{m}+2\pi i\, {}^t\vec{m}(\vec{z}+\vec{a}_i)])}{\exp(-\sum_i [\pi i\, {}^t\vec{m}\Omega\vec{m}+2\pi i\, {}^t\vec{m}(\vec{z}+\vec{b}_i)])}\, f(\vec{z}) = f(\vec{z}).$$

A variation of this method uses theta functions with characteristic: If $\vec{a},\vec{b},\vec{a}',\vec{b}' \in \frac{1}{N}\mathbb{Z}^g$, then

$$\frac{\vartheta[{\vec{a} \atop \vec{b}}](N\vec{z},\Omega)}{\vartheta[{\vec{a} \atop \vec{b}}](N\vec{z},\Omega)} \quad \text{is a meromorphic function on } X_\Omega.$$

Likewise, if $\sum a_i \equiv \sum a_i'$, $\sum b_i \equiv \sum b_i'$ mod $\mathbb{Z}^n$, then

$$\frac{\prod_i \vartheta\begin{bmatrix}\vec{a}_i\\ \vec{b}_i\end{bmatrix}(\vec{z},\Omega)}{\prod_i \vartheta\begin{bmatrix}\vec{a}_i'\\ \vec{b}_i'\end{bmatrix}(\vec{z},\Omega)} \quad \text{is a meromorphic function on } X_\Omega.$$

(the 1st one because

$$\frac{\vartheta[{a \atop b}](N(z+m+\Omega m'),\Omega)}{\vartheta[{a' \atop b'}](N(z+m+\Omega m'),\Omega)} =$$

$$= \frac{\exp(2\pi i\, {}^t aNm)\exp(-2\pi i\, {}^t bNm')\,[\exp(-\pi i N^2\, {}^t m'\Omega m'-2\pi i N\, {}^t m'z)]\,\vartheta[{a \atop b}](Nz,\Omega)}{\exp(2\pi i\, {}^t a'Nm)\exp(-2\pi i\, {}^t b'Nm')\,[\exp(-\pi i N^2\, {}^t m'\Omega m'-2\pi i N\, {}^t m'z)]\,\vartheta[{a' \atop b'}](Nz,\Omega)}$$

and $(a-a')N$, $(b-b')N \in \mathbb{Z}^g$; similarly the 2nd one).

Method II:

$$\left[\frac{\frac{\partial\vartheta}{\partial z_i}(\vec{z}+\vec{a},\Omega)}{\vartheta(\vec{z}+\vec{a},\Omega)} - \frac{\frac{\partial\vartheta}{\partial z_i}(\vec{z}+\vec{b},\Omega)}{\vartheta(\vec{z}+\vec{b},\Omega)}\right] = \frac{\partial}{\partial x_i}\log\left(\frac{\vartheta(\vec{z}+\vec{a},\Omega)}{\vartheta(\vec{z}+\vec{b},\Omega)}\right)$$

is a meromorphic function on $X_\Omega$ as is $\frac{\partial}{\partial z_i}\log\left(\frac{\vartheta[{a \atop b}](z,\Omega)}{\vartheta[{a' \atop b'}](z,\Omega)}\right)$

(because the ratio of the 2 $\vartheta$'s is multiplied by a constant when z is replaced by $z+\Omega n+m$).

Method III:

$$\frac{\vartheta\,\frac{\partial^2\vartheta}{\partial z_i\partial z_j} - \frac{\partial\vartheta}{\partial z_i}\,\frac{\partial\vartheta}{\partial z_j}}{\vartheta^2} = \frac{\partial^2}{\partial z_i\partial z_j}\log\vartheta$$

is a meromorphic function on $X_\Omega$

(it increases by $\frac{\partial^2}{\partial z_i\partial z_j}\log\exp(-\pi i\,{}^t m\Omega m-2\pi i\,{}^t mz)$ when $z \longmapsto z+\Omega m+m'$ and this is zero).
This is the Weierstrass $\wp$-function when $g = 1$.

Now let $\Omega$ be the period matrix of the compact Riemann Surface X again. The applications of $\vartheta$ to the function theory on X are based on a fundamental result of Riemann who computed the zeroes of

$$f(P) = \vartheta\left(\vec{z} + \int_{P_0}^{P} \vec{\omega},\Omega\right) \qquad (\vec{z}\text{ fixed}).$$

Note that f(P) is a locally single-valued but globally multivalued function, which is invariant around the A-periods, but, on prolongation around a B-period $B_k$, f is multiplied by

$$\exp[-\pi i\Omega_{kk} - 2\pi i(\int_{P_0}^{P}\omega_k + z_k)].$$

<u>Theorem</u> 3.1 (<u>Riemann</u>): <u>There is a vector</u> $\vec{\Delta} \in \mathbb{C}^g$, <u>such that for all</u> $\vec{z} \in \mathbb{C}^g$, $f(\vec{z}) = \vartheta(\vec{z} + \int_{P_0}^{P} \vec{\omega}, \Omega)$ <u>either vanishes identically, or has</u> g <u>zeroes</u> $Q_1, \cdots, Q_g$ <u>such that</u>

$$\sum_{i=1}^{g} \int_{P_0}^{Q_i} \vec{\omega} \equiv -\vec{z} + \vec{\Delta} \pmod{L_\Omega}.$$

<u>Proof</u>: This is another application of Green's Theorem. We cut open the Riemann Surface X as before. We may assume that $Q_i \in X_0$ and $P_0 \in X_0$. Let $\Delta_i$ be a small disc around $Q_i$. We consider the 1-form df/f on $X_0 - \cup\Delta_i$. It is holomorphic, hence closed, so

$$0 = \int_{X_0 - \cup\Delta_i} d\left(\frac{df}{f}\right)$$

$$= \int_{\partial(X_0 - \cup\Delta_i)} \frac{df}{f}$$

$$= -\sum_i \int_{\partial\Delta_i} \frac{df}{f} + \sum_{k=1}^{g} \int_{(A_k^- - A_k^+)} \frac{df}{f} + \sum_{k=1}^{g} \int_{(B_k^- - B_k^+)} \frac{df}{f}.$$

Now f is invariant under the A-periods, hence it has the same value on $B_k^+, B_k^-$. And f increased by $-2\pi i \omega_k$ on $B_k$ which joins $A_k^-$ to $A_k^+$.

Thus the middle integral equals

$$\sum_{k=1}^{g} 2\pi i \int_{A_k} \omega_k = 2\pi i g$$

and the last integral is zero. Since $\int_{\partial\Delta_i} \frac{df}{f} = 2\pi i(\text{mult. of zero } Q_i)$,

this proves that the number of zeroes of f is exactly g (counted with multiplicity if necessary).

Next, let $\omega_k = dg_k$ with $g_k(P_o) = 0$ on $X_0$ and repeat the same argument with the 1-form $g_k \cdot \frac{df}{f}$:

$$0 = \int_{X_0 - \cup\Delta_i} d(g_k \frac{df}{f})$$

$$= -\sum_{i=1}^{g} \int_{\partial\Delta_i} g_k \frac{df}{f} + \sum_{\ell=1}^{g} \int_{(A_\ell^- - A_\ell^+)} g_k \frac{df}{f} + \sum_{\ell=1}^{g} \int_{(B_\ell^- - B_\ell^+)} g_k \frac{df}{f} .$$

Taking these terms one at a time:

$$\int_{\partial\Delta_i} g_k \frac{df}{f} = 2\pi i \; g_k(Q_i) = 2\pi i \int_{P_0}^{Q_i} \omega_k .$$

Next $g_k$ on $B_\ell^-$ is $g_k$ on $B_\ell^+$ plus $\delta_{k\ell}$ because the path $A_\ell$ leads from $B_\ell^+$ to $B_\ell^-$. So

$$\int_{(B_\ell^- - B_\ell^+)} g_k \frac{df}{f} = \delta_{k\ell} \int_{B_\ell} \frac{df}{f}$$

$$= \delta_{k\ell} \text{ (change in value of log f around } B_\ell)$$

$$= \delta_{k\ell} \left[-\pi i \Omega_{\ell\ell} - 2\pi i \int_{P_0}^{P_1} \omega_\ell - 2\pi i z_\ell + 2\pi i \underset{m_\ell}{\left(\begin{matrix}\text{some}\\ \text{integer}\end{matrix}\right)}\right]$$

(where $P_1$ is the base point of the paths $B_\ell$ and $A_\ell$).

Next $g_k$ on $A_\ell^+$ is $g_k$ on $A_\ell^-$ plus $(-\Omega_{k\ell})$ because the path $B_\ell$ leads from $A_k^-$ to $A_k^+$. So

$$\int_{(A_\ell^- - A_\ell^+)} g_k \frac{df}{f} = \int_{A_\ell^+} (g_k - \Omega_{k\ell})\left(\frac{df}{f} - 2\pi i\omega_\ell\right) - \left(g_k \cdot \frac{df}{f}\right)$$

$$= -\Omega_{k\ell} \int_{A_\ell^+} \frac{df}{f} - 2\pi i \int_{A_\ell^+} g_k\omega_\ell + 2\pi i\Omega_{k\ell} \int_{A_\ell} \omega_\ell$$

$$= +\Omega_{k\ell}\, 2\pi i \begin{pmatrix} \text{some} \\ \text{integer} \\ n_\ell \end{pmatrix} - 2\pi i \int_{A_\ell^+} g_k\omega_\ell + 2\pi i\Omega_{k\ell}.$$

Putting all this together, we find:

$$\sum_{i=1}^{g} \int_{P_0}^{Q_i} \omega_k = -z_k + \left[-\frac{\Omega_{kk}}{2} - \int_{P_0}^{P_1} \omega_k + \sum_\ell \Omega_{k\ell} - \sum_\ell \int_{A_\ell^+} g_k\omega_\ell\right]$$

$$+ \left(m_k + \sum_\ell \Omega_{k\ell} n_\ell\right)$$

which proves the theorem. QED

To exploit this theorem, we make a few definitions: let $\mathrm{Symm}\,^n X$ be the compact analytic space constructed by dividing $X \times \cdots \times X$ (n factors) by the action of the permutation group in n letters, permuting the factors:

$$\mathrm{Symm}\,^n X = \overbrace{X \times \cdots \times X}^{n\times} / \mathfrak{S}_n$$

This is well-known to be a manifold, even at points where $\mathfrak{S}_n$ is not acting freely. In fact, if

$$(x_1,\cdots,x_n) \in X^n$$

and

$$x_1 = \cdots = x_k = P, \quad x_{k+1},\cdots,x_n \neq P.$$

(this can always be achieved by permuting the $x_i$'s), then let U be a neighborhood of P disjoint from an open subset $X_0 \subset X$ containing $x_{k+1},\cdots,x_n$. Then $\mathrm{Symm}^n X$ contains the open set:

$$(U^k/\mathfrak{S}_k) \times (X_0^{n-k}/\mathfrak{S}_{n-k}) .$$

Let $\zeta$ be a coordinate on U. Then coordinates on $U^k/\mathfrak{S}_k$ are given by the elementary symmetric functions in $\zeta(x_1),\cdots,\zeta(x_k)$, hence $U^k/\mathfrak{S}_k$ is an open subset of $\mathbb{C}^k$. By induction this proves that $\mathrm{Symm}^n X$ is a manifold.

Points of $\mathrm{Symm}^n X$ are in 1-1 correspondence with so-called "divisors" on X, positive and of degree n. These are finite formal sums:

$$\sum_{i=1}^{\ell} k_i P_i$$

of points of X, with

$$k_i > 0, \quad \sum_{i=1}^{\ell} k_i = n.$$

We will usually write divisors as:

$$\sum_{i=1}^{n} P_i$$

allowing the $P_i$'s to be equal. As in §2, we have a canonical map

$$I_n: \ \mathrm{Symm}^{\,n}X \longrightarrow \mathrm{Jac}(X)$$

given by

$$I_n(\sum P_i) = \left(\sum_{i=1}^{n} \int_{P_0}^{P_i} \vec{\omega} \mod L_\Omega\right).$$

Clearly, $I_n$ is a holomorphic map from $\mathrm{Symm}^{\,n}X$ to Jac(X).

To exploit Riemann's theorem, define $\mathcal{E} \subset \mathrm{Jac}(X)$ to be the proper closed analytic subset:

$$\mathcal{E} = \left\{\vec{z} \,\middle|\, \vartheta\left(\vec{\Delta}-\vec{z} + \int_{P_0}^{P} \vec{\omega}\right) = 0, \text{ all } P \in X\right\}.$$

Let $U = \mathrm{Jac}(X) - \mathcal{E}$. Then I claim:

<u>Corollary</u> 3.2: <u>For all</u> $P_1,\cdots,P_g \in X$, $\vec{z} \in \mathbb{C}^g$:

$$\sum_{i=1}^{g} \int_{P_0}^{P_i} \vec{\omega} \equiv \vec{z} \mod L_\Omega \Longrightarrow \vartheta\left(\vec{\Delta}-\vec{z} + \int_{P_0}^{P_i} \vec{\omega}\right) = 0 \ \underline{\text{for all}}\ i$$

<u>and if</u> $\vec{z} \notin \mathcal{E}$, <u>then the divisor</u> $\sum_{1}^{g} P_i$ <u>is uniquely determined by</u> $\vec{z}$ <u>by</u>:

$$\sum_{i=1}^{g} \int_{P_0}^{P_i} \vec{\omega} \equiv \vec{z} \mod L_\Omega .$$

*Hence*

$$I_g: \ \text{Symm}^g X \longrightarrow \text{Jac}(X)$$

*is bimeromorphic. More precisely, it is surjective and*

$$\text{res } I_g: \ I_g^{-1}(U) \longrightarrow U$$

*is an isomorphism.*

*Proof*: Let $W \subset \text{Symm}^g X \times \text{Jac}(X)$ be the closed analytic subset defined by *both* conditions

$$\sum \int_{P_0}^{P_i} \vec{\omega} \equiv \vec{z} \bmod L_\Omega$$

and

$$f(P) = \vartheta\left(\vec{\Delta} - \vec{z} + \int_{P_0}^{P} \vec{\omega}\right)$$

is zero on $\sum_{i=1}^{g} P_i$. Consider the projections

$$\begin{array}{ccc} & W & \\ {\scriptstyle p_1}\swarrow & & \searrow{\scriptstyle p_2} \\ \text{Symm}^g X & & \text{Jac}(X) \end{array}$$

By Riemann's theorem, $p_2^{-1}(U) \longrightarrow U$ is an isomorphism. In particular, $p_2(W)$ is a closed subset of $\text{Jac}(X)$ containing $U$, hence equals $\text{Jac}(X)$. Thus $\dim W \geq g$. But $p_1$ is injective because the $P_i$ determine $\vec{z}$. So $\dim p_1(W) \geq g$, hence $p_1$ is surjective, hence $p_1$ is bijective. Therefore $W$ is nothing but the graph of $I_g$, i.e., the first condition implies the second. This is the first assertion of the Corollary and the rest is a restatement of Riemann's theorem. QED

We next investigate the function

$$E_{\vec{e}}(x,y) = \vartheta\left(\vec{e} + \int_x^y \vec{\omega}\right)$$

where $\vec{e} \in \mathbb{C}^g$ is fixed and satisfies $\vartheta(\vec{e}) = 0$, and $x,y \in X$. As with f, $E_{\vec{e}}$ is locally single-valued, but globally multi-valued, being multiplied by an exponential factor when x or y are carried around a B-period.

Lemma 3.3. For any $P \in X$,

$$D_P = \left\{\vec{e} \in \mathbb{C}^g/L_\Omega \,\middle|\, \vartheta\left(\vec{e} + \int_P^y \vec{\omega}\right) = 0, \text{ all } y \in X\right\}$$

is an analytic subset of Jac(X) of codimension at least 2. Hence for any finite subset $P_1,\cdots,P_n$ of X, there is an $\vec{e}$ such that

$$\vartheta(\vec{e}) = 0, \quad f_i(y) = \vartheta\left(\vec{e} + \int_{P_i}^y \vec{\omega}\right) \not\equiv 0 \quad \text{for all } i.$$

Proof: Let D be an irreducible component of $D_P$ and let $X_P \subset \text{Jac}(X)$ be the locus of points $\int_P^y \vec{\omega}$, all $y \in X$. Consider the locus of points a+b, $a \in D$, $b \in X_P$ and call it $D+X_P$. Then $D+X_P$ is an irreducible analytic subset of Jac(X) containing $D_P$ and contained in the locus of zeroes of $\vartheta$. Hence $\dim D + X_P \leq g-1$. If $\dim D = g-1$, it follows that $\dim D = \dim(D+X_P)$, hence $D = (D+X_P)$. But then $D + X_P + X_P + \cdots + X_P = D$. By the

Corollary, $I_g$ is surjective, i.e.,

$$\overbrace{X_P + \cdots + X_P}^{g\times} = \mathrm{Jac}(X).$$

Together these imply $D = \mathrm{Jac}(X)$, which is a contradiction. Thus $\dim D \leq g-2$. QED

Lemma 3.4: Let $\vec{e} \in \mathbb{C}^g$ satisfy $\vartheta(\vec{e}) = 0$, $E_{\vec{e}}(x,y) \not\equiv 0$. Then there are $2g-2$ points $R_1,\cdots,R_{g-1},S_1,\cdots,S_{g-1} \in X$ such that

$$E_{\vec{e}}(x,y) = 0 \iff \begin{array}{ll} & \text{a)}\ x = y \\ \text{or} & \text{b)}\ x = R_i \\ \text{or} & \text{c)}\ y = S_i\ . \end{array}$$

More precisely, including multiplicities, the divisor of zeroes of $E_{\vec{e}}$ is the sum of

a) $\Delta$, the diagonal

b) $\{R_i\} \times X$, $1 \leq i \leq g-1$

c) $X \times \{S_i\}$, $1 \leq i \leq g-1$.

Proof: Let $R \in X$ be any point such that $E_{\vec{e}}(R,y) \not\equiv 0$. Then by Riemann's theorem, there are $g$ points $y_1,\cdots,y_g$ such that $\vartheta(\vec{e} + \int_R^{y_i} \vec{\omega}) = 0$ and by the Corollary:

$$(y_1,\cdots,y_g) = \left\{ \begin{array}{l} \text{unique unordered g-tuple such that} \\ \sum_{i=1}^{g} \int_{P_0}^{y_i} \vec{\omega} \equiv \vec{\Delta} - \left(\vec{e} + \int_R^{P_0} \vec{\omega}\right) \bmod L_\Omega \end{array} \right\}$$

Since $\vartheta(\vec{e} + \int_R^R \vec{\omega}) = 0$, we may assume $y_1 = R$. It follows that:

$$(y_2,\cdots,y_g) = \left\{ \begin{array}{l} \text{unique unordered (g-1)-tuple such that} \\ \sum_{i=2}^{g} \int_{P_0}^{y_i} \vec{\omega} \equiv \vec{\Delta} - \vec{e} \end{array} \right\} .$$

Therefore $y_2,\cdots,y_g$ depend only on $\vec{e}$: set $S_i = y_{i-1}$.

There is a finite set of points $R \in X$ such that $\vartheta(\vec{e} + \int_R^y \vec{\omega}) \equiv 0$.

To investigate their number, choose $S_0 \neq S_1,\cdots,S_{g-1}$. Then

$$\vartheta(\vec{e} + \int_x^{S_0} \vec{\omega}) = 0 \quad \text{if and only if} \quad x = S_0 \text{ or } \vartheta(\vec{e} + \int_x^y \vec{\omega}) = 0, \text{ all } y.$$

But $\vartheta(-\vec{z}) = \vartheta(\vec{z})$, so

$$\vartheta(\vec{e} + \int_x^{S_0} \vec{\omega}) = \vartheta(-\vec{e} + \int_{S_0}^x \vec{\omega})$$

has g zeroes by Riemann's theorem. So there are g-1 points $R_i \in X$ such that $\vartheta(\vec{e} + \int_{R_i}^y \vec{\omega}) \equiv 0$. <u>QED</u>

With these functions $E_{\vec{e}}$, we can now prove Abel's Theorem (see §2): given cycles $\sum_{i=1}^{d} P_i$, $\sum_{i=1}^{d} Q_i$ of the same degree, assume

$$I_d(\sum P_i) = I_d(\sum Q_i).$$

Then we want to construct a meromorphic function on X with poles $\sum P_i$, zeroes $\sum Q_i$. In fact, we choose $\vec{e} \in \mathbb{C}^g$ so that

$$\vartheta(\vec{e}) = 0$$

$$E_{\vec{e}}(P_i,y) \not\equiv 0$$

$$E_{\vec{e}}(Q_i,y) \not\equiv 0.$$

Consider the function on X:

$$f(y) = \frac{\prod_{i=1}^{d} E_{\vec{e}}(Q_i,y)}{\prod_{i=1}^{d} E_{\vec{e}}(P_i,y)} . \tag{3.5}$$

We fix the sheets of the multivalued functions $E_{\vec{e}}$ a little more precisely as follows:

Choose paths $\sigma_i$ from $P_0$ to $P_i$, $\tau_i$ from $P_0$ to $Q_i$ so that

$$\sum_{i=1}^{d} \int_{\sigma_i} \vec{\omega} = \sum_{i=1}^{d} \int_{\tau_i} \vec{\omega} .$$

Note that for any $\sigma_i, \tau_i$, by our assumption $I_d(\sum P_i) = I_d(\sum Q_i)$, the above sums would be congruent mod $L_\Omega$. Hence altering one of them, we can achieve equality in $\mathbb{C}^g$.

Define f as:

$$f(y) = \frac{\prod_{i=1}^{d} \vartheta(\vec{e} + \int_{Q_i}^{y} \vec{\omega})}{\prod_{i=1}^{d} \vartheta(\vec{e} + \int_{P_i}^{y} \vec{\omega})}$$

where the paths from $Q_i$, resp. $P_i$, to y are $-\tau_i$, resp. $-\sigma_i$, followed by the same path from $P_0$ to y in all integrals. If we do this, let us examine the effect of moving y around a path in X or of altering the path from $P_0$ to y. If the change is by $A_k$, nothing happens. If the change is by $B_k$, f is multiplied by:

$$\frac{\prod_{i=1}^{d} \exp\left[-\pi i \Omega_{kk} - 2\pi i (\int_{Q_i}^{y} \omega_k + e_k)\right]}{\prod_{i=1}^{d} \exp\left[-\pi i \Omega_{kk} - 2\pi i (\int_{P_i}^{y} \omega_k + e_k)\right]}$$

$$= \exp\left[-2\pi i (\sum_{i=1}^{d} \int_{Q_i}^{y} \omega_k - \sum_{i=1}^{d} \int_{P_i}^{y} \omega_k)\right]$$

$$= 1.$$

Thus f is single-valued. By Lemma 3.4, its zeroes are precisely the $Q_i$ and its poles the $P_i$. This proves Abel's Theorem.

The beautiful function $E_{\vec{e}}$ plays the role for X of the function x-y for the rational Riemann Surface $\mathbb{P}^1$: namely on $\mathbb{P}^1$, every rational function can be factored

$$f(y) = c \frac{\Pi(y-Q_i)}{\Pi(y-P_i)} .$$

(3.5) is a generalization of the formula to all compact Riemann surfaces. $E_{\vec{e}}$ is called the "Prime form" because of this role in factoring meromorphic functions.

We conclude the section with one last consequence of Riemann's theorem:

<u>Corollary</u> 3.6: <u>For all</u> $\vec{e} \in \mathbb{C}^g$

$$\vartheta(\vec{e}) = 0 \iff \left[ \exists P_1,\cdots,P_{g-1} \in X \ \underline{\text{such that}} \quad \vec{e} = \vec{\Delta} - \sum_{i=1}^{g-1} \int_{P_0}^{P_i} \vec{\omega} \right]$$

<u>Proof</u>: $\Leftarrow$ In fact take any $P_g \in X$ and apply (3.2) to $\vec{z} = \sum_1^g \int_{P_0}^{P_i} \vec{\omega}$. It follows that

$$\vartheta\left(\vec{\Delta} - \sum_{i=1}^{g-1} \int_{P_0}^{P_i} \vec{\omega}\right) = 0 .$$

Conversely, to prove "$\Longrightarrow$" note that $\vartheta(\vec{e}) = 0$ defines a codimension 1 subset of Jac(X) and the right hand side defines a closed analytic subset of Jac(X). Moreover by (3.3),

$$\left\{ \vec{e} \;\middle|\; \vartheta(\vec{e} + \int_{P_0}^{y} \vec{\omega}) = 0, \quad \text{all} \quad y \in X \right\}$$

has codimension 2. So if we prove "$\Longrightarrow$" for $\vec{e}$'s such that

$$\vartheta(\vec{e}) = 0$$

$$\exists \quad y \in X, \quad \vartheta(\vec{e} + \int_{P_0}^{y} \vec{\omega}) \not\equiv 0,$$

it follows for all $\vec{e}$. Take such an $\vec{e}$. By the surjectivity of $I_g$, we can write it as:

$$\vec{e} = \vec{\Delta} - \sum_{i=1}^{g} \int_{P_0}^{P_i} \vec{\omega}.$$

Consider

$$f(y) = \vartheta\left(\vec{e} + \int_{P_0}^{y} \vec{\omega}\right).$$

By (3.2), the divisor of zeroes of f(y) is exactly $\sum_{i=1}^{g} P_i$. On the other hand, $\vartheta(\vec{e}) = 0$ so $P_0$ is a zero of f(y). Therefore some $P_i$ equals $P_0$, and $\vec{e}$ has the form required by the Corollary.

QED

## Appendix to §3: The meaning of $\vec{\Delta}$

In the preceding discussion, $\vec{\Delta}$ comes up as a strange bi-product of an elaborate Green's theorem calculation. We would like to apply the Riemann-Roch theorem on X to give another point of view on the Corollaries to Riemann's Theorem (3.1), leading to a determination of $\vec{\Delta}$ from another point of view. We need some of the standard terminology connected with divisors on X:

Definition 3.7: 2 divisors $D_1, D_2$ on X are linearly equivalent if equivalently $I(D_1-D_2) = 0$ or $\exists$ a meromorphic function f such that $D_1-D_2$ = (zeroes of f) - (poles of f). This is written $D_1 \equiv D_2$. An equivalence class of linearly equivalent divisors is called a divisor class. Under +, the set of divisor classes is a group, called Pic X (thus the Jacobian Jac X is isomorphic to the subgroup of Pic X of divisor classes of degree 0).

Definition 3.8: Let $\omega$ be any meromorphic 1-form on X. For all $P \in X$, let z be a local coordinate on X near P and write

$$\omega = z^{n_P}\cdot u(z)\cdot dz$$

where $n_P \in \mathbb{Z}$, u(z) is holomorphic and $u(0) \neq 0$. Then the divisor of $\omega$ is:

$$(\omega) = \sum_{P\in X} n_P \cdot P.$$

Note that if $\omega_1, \omega_2$ are 2 such 1-forms, $\omega_1/\omega_2 = f$, a meromorphic function on X, so

$$(\omega_1)-(\omega_2) = \text{divisor of } f \equiv 0.$$

Thus these divisors $(\omega)$ all lie in the same divisor class $K_X$, called the canonical divisor class on X.

Definition 3.9: The set $\Sigma$ of divisor classes D such that

$$2D \equiv K_X$$

is called the set of theta characteristics of X.

Note that $\Sigma$ is a principal homogeneous space under $(\text{Pic } X)_2$, the group of 2-torsion in Pic X (i.e., $\forall\, D_1, D_2 \in \Sigma$, there is a unique $E \in (\text{Pic } X)_2$ such that $D_1 = D_2 + E$), hence card. $\Sigma$ = card. $(\text{Pic } X)_2$. Moreover, all $D \in (\text{Pic } X)_2$ have degree 0, so

$$(\text{Pic } X)_2 = (\text{Jac } X)_2 = \tfrac{1}{2}L_\Omega / L_\Omega \cong (\mathbb{Z}/2\mathbb{Z})^{2g},$$

and both $\Sigma, (\text{Pic } X)_2$ have $2^{2g}$ elements.

The main result of this appendix is:

Theorem 3.10: Let $\Theta \subset \mathbb{C}^g/L_\Omega$ be the analytic subset defined by $\vartheta(\vec{z},\Omega) = 0$. We consider all translates $\vec{e}+\Theta$ of the subset $\Theta$ by a point of $\mathbb{C}^g/L_\Omega$. Then:

a) The map

$$D \longmapsto \left(\begin{array}{c}\text{locus of points } I(P_1+\ldots+P_{g-1}-D) \\ \text{for all } P_1,\cdots,P_{g-1} \in X\end{array}\right) \subset \mathbb{C}^g/L_\Omega$$

is an isomorphism

$$\Sigma \xrightarrow{\approx} \left(\begin{array}{c}\text{set of translates } \vec{e}+\Theta \text{ which are} \\ \text{symmetric, i.e., invariant under} \\ \vec{z} \longmapsto -\vec{z}\end{array}\right).$$

b) The map

$$\binom{\eta'}{\eta''} \longmapsto \left(\frac{\text{locus of zeroes of}}{\vartheta[\begin{smallmatrix}\eta'\\\eta''\end{smallmatrix}](\vec{z},\Omega)}\right) \subset \mathbb{C}^g/L_\Omega$$

is an isomorphism

$$\tfrac{1}{2}\mathbb{Z}^{2g}/\mathbb{Z}^{2g} \xrightarrow{\approx} \left(\begin{array}{l}\text{set of translates } \vec{e}+\Theta \text{ which}\\ \text{are symmetric}\end{array}\right).$$

Proof: Note that $\vartheta(-\vec{z}) = \vartheta(\vec{z})$ (this is immediate from the formula defining $\vartheta$), so $\Theta$ itself is symmetric. But $\vartheta[\begin{smallmatrix}\eta'\\\eta''\end{smallmatrix}](\vec{z}) = 0$ is the translate of $\Theta$ by $\Omega\eta'+\eta''$. If $\eta',\eta'' \in \frac{1}{2}\mathbb{Z}^g$, then $\Omega\eta'+\eta'' \in \mathbb{C}^g/L_\Omega$ is of order 2 and a translate of a symmetric subset of a group by an element of order 2 is symmetric. In fact, a translate of a symmetric subset $\Theta$ by an element a is also symmetric if and only if $\Theta$ is invariant by translation by 2a:

$$-(\Theta+a) = \Theta+a \iff \Theta+2a = \Theta .$$

Thus to prove (b), we need to check that $\Theta \neq \Theta+\vec{e}$, for all $\vec{e} \neq 0$. This is proven similarly to Lemma 1.7: say $\vec{e} \in \mathbb{C}^g$ satisfies:

$$\vartheta(\vec{z}) = 0 \iff \vartheta(\vec{z}+\vec{e}) = 0$$

and we must show $\vec{e} \in L_\Omega$. Consider $\frac{\vartheta(\vec{z}+\vec{e})}{\vartheta(\vec{z})}$: the zeroes of numerator and denominator cancel out*, hence this is a nowhere zero holomorphic function on $\mathbb{C}^g$. If f(z) is its logarithm, we have

*Note that by (3.6), $\vartheta(\vec{z}) = 0$ is an irreducible analytic subset of $\mathbb{C}^g/L_\Omega$ and that by (3.1), $\vartheta$ vanishes to 1st order on it: otherwise the $f(\vec{z})$ in (3.1) would always have multiple roots and by (3.2), we can take the $P_i$'s distinct in general position.

$$\vartheta(\vec{z}+\vec{e}) = \exp f(\vec{z})\cdot\vartheta(\vec{z}).$$

In this substitute $\vec{z}+\Omega\vec{n}+\vec{m}$ for $\vec{z}$ and use the functional equation for $\vartheta$ on both sides. We find that

$$\exp(-\pi i\,{}^t n\Omega n-2\pi i\,{}^t n.(z+e) + f(z)).\vartheta(z)$$

$$= \exp(f(z+\Omega n+m) - \pi i\,{}^t n\Omega n-2\pi i\,{}^t nz)\,\vartheta(z)$$

hence

$$(*) \qquad f(z+\Omega n+m) - f(z) = -2\pi i\,{}^t ne + 2\pi i\cdot(\text{integer})$$

for all $\vec{z} \in \mathbb{C}^g$. Therefore, $\partial f/\partial z_i$ is invariant by $z \longmapsto z+\Omega n+m$, i.e., is a holomorphic function on $\mathbb{C}^g/L_\Omega$. But then $\partial f/\partial z_i$ must be a constant, hence f is a linear function. If $f(z) = c_o + 2\pi i\,{}^t\vec{c}\cdot\vec{z}$, then (*) says

$$ {}^t\vec{c}\Omega\cdot\vec{n} = -{}^t n\cdot e + (\text{integer}), \qquad \text{all } \vec{n} \in \mathbb{Z}^g$$

$$ {}^t\vec{c}\cdot\vec{m} = (\text{integer}), \qquad \text{all } \vec{m} \in \mathbb{Z}^g .$$

By the 2nd formula, $\vec{c} \in \mathbb{Z}^g$, hence by 1st

$$\vec{e} = -\Omega\cdot\vec{c} + (\text{integral vector}) \in L_\Omega.$$

This proves (b).

To prove (a), we use the following consequence of the Riemann-Roch theorem: For all divisors E of degree g-1,

$$\begin{bmatrix} \exists P_1,\cdots,P_{g-1} \in X \text{ s.t.} \\ E \equiv \sum_1^{g-1} P_i \end{bmatrix} \iff \begin{bmatrix} \exists Q_1,\cdots,Q_{g-1} \in X \text{ s.t.} \\ K_X - E \equiv \sum_1^{g-1} Q_i \end{bmatrix}$$

Therefore if $D \in \Sigma$, $\forall P_1,\cdots,P_{g-1} \in X$, $\exists Q_1,\cdots,Q_{g-1} \in X$ such that

$$2D - (P_1+\cdots+P_{g-1}) \equiv Q_1+\cdots+Q_{g-1}$$

or

$$I(P_1+\cdots+P_{g-1}-D) = -I(Q_1+\cdots+Q_{g-1}-D).$$

This proves the symmetry of the locus in (a). By (3.6), this locus is a translate of $\Theta$. Finally, $\Sigma$ has $2^{2g}$ elements in it and by part (b), there are $2^{2g}$ symmetric $(\Theta+\vec{e})$'s. This proves (a).

QED

Corollary 3.11: $\vec{\Delta} = I(D_o-(g-1)P_o)$ for some $D_o \in \Sigma$.

Proof: Let $D_o \in \Sigma$ map under (a) to $\Theta$ itself. Compare 3.6 and 3.10: it follows that

$$\left\{\text{locus of pts } I(P_1+\cdots+P_{g-1}-D_o)\right\} = \left\{\text{locus of pts } I(P_1+\cdots+P_{g-1}-(g-1)P_o)-\vec{\Delta}\right\}$$

Therefore translation by $\vec{\Delta}+I((g-1)P_o-D_o)$ carries $\Theta$ to itself, hence is 0. QED

Corollary 3.12: Let $D \in \Sigma$ and $\binom{\eta'}{\eta''} \in \frac{1}{2}\mathbb{Z}^{2g}/\mathbb{Z}^{2g}$ correspond to the same symmetric translate of $\Theta$ in (3.10). Then

$$\left(\vartheta[{}^{\eta'}_{\eta''}](0,\Omega)=0\right) \Longleftrightarrow \left(\vartheta(\Omega\eta'+\eta'',\Omega)=0\right) \Longleftrightarrow \begin{pmatrix} \exists P_1,\cdots,P_{g-1} \in X \text{ such that} \\ D \equiv \sum P_i \end{pmatrix}$$

<u>Proof</u>: Immediate from 3.10.

The set of theta characteristics has an important 2-valued form defined on it:

<u>Definition</u> 3.13: <u>For all</u> $\zeta = ({}^{\zeta'}_{\zeta''}) \in \frac{1}{2}\mathbb{Z}^{2g}/\mathbb{Z}^{2g}$, <u>define</u>

$$e_*(\zeta) = (-1)^{4\,{}^t\zeta'\cdot\zeta''} .$$

<u>For all</u> $\zeta = ({}^{\zeta'}_{\zeta''})$, $\eta = ({}^{\eta'}_{\eta''}) \in \frac{1}{2}\mathbb{Z}^{2g}/\mathbb{Z}^{2g}$, <u>define</u>

$$e_2(\zeta,\eta) = (-1)^{4({}^t\zeta'\cdot\eta''-{}^t\eta'\cdot\zeta'')} .$$

We want to think of $e_*$ as the exponential of a quadratic form on $(\mathbb{Z}/2\mathbb{Z})^{2g}$ with values in $\mathbb{Z}/2\mathbb{Z}$, and of $e_2$ as the exponential of a skew-symmetric form $(\mathbb{Z}/2\mathbb{Z})^{2g}\times(\mathbb{Z}/2\mathbb{Z})^{2g} \longrightarrow \mathbb{Z}/2\mathbb{Z}$. Since $+1 = -1$ in $\mathbb{Z}/2\mathbb{Z}$, a skew-symmetric form is also symmetric and these 2 are related by:

$$\frac{e_*(\zeta+\eta)}{e_*(\zeta)\cdot e_*(\eta)} = e_2(\zeta,\eta) .$$

The importance of $e_*$ rests on

<u>Proposition</u> 3.14: <u>For all</u> $\zeta \in \frac{1}{2}\mathbb{Z}^{2g}/\mathbb{Z}^{2g}$

$$\vartheta[{}^{\zeta'}_{\zeta''}](-\vec{z},\Omega) = e_*(\zeta)\cdot\vartheta[{}^{\zeta'}_{\zeta''}](\vec{z},\Omega).$$

Proof: This is an easy calculation:

$$\vartheta[{\zeta' \atop \zeta''}](-\vec{z},\Omega) = \sum_{n\in\mathbb{Z}^g} \exp[\pi i\,{}^t(n+\zeta')\Omega(n+\zeta')+2\pi i\,{}^t(n+\zeta')(-z+\zeta'')]$$

$$= \sum_{m\in\mathbb{Z}^g} \exp[\pi i\,{}^t(m+\zeta')\Omega(m+\zeta')+2\pi i\,{}^t(m+\zeta')(z-\zeta'')]$$

$$\text{if} \quad m = -n - 2\zeta'$$

$$= \exp(4\pi i\,{}^t\zeta'.\zeta'')\cdot\vartheta[{\zeta' \atop \zeta''}](\vec{z},\Omega). \qquad \text{QED}$$

Corollary 3.15: For all $\zeta \in \frac{1}{2}\mathbb{Z}^{2g}/\mathbb{Z}^{2g}$

$e_*(\zeta) = +1 \iff$ (the divisor $\vartheta[\zeta](z,\Omega)=0$ does not contain 0, or else has a point of even mult. at 0) $\iff$ ($\Theta$ does not contain $\Omega\eta'+\eta''$ or else has a point of even mult. there)

$e_*(\zeta) = -1 \iff$ (the divisor $\vartheta[\zeta](z,\Omega)=0$ contains 0 and has a point of odd mult. at 0) $\iff$ ($\Theta$ contains $\Omega\eta'+\eta''$ and has a point of odd mult. there)

Here a divisor D on a complex manifold is a locus defined by one equation $f(z_1,\cdot\cdot,z_n) = 0$ in local coordinates, and 0 is a point of multiplicity k if all terms $(az_1^{k_1}z_2^{k_2}\cdots z_n^{k_n})$ in f of total degree $k_1 +\cdot\cdot+ k_n < k$ vanish while at least one term in f of total degree $k_1 +\cdot\cdot+ k_n = k$ does not vanish. The Corollary comes from the fact that if $f(-z) = f(z)$, all terms in f have even total degree, while if $f(-z) = -f(z)$, all terms in f have odd total degree. The Corollary shows that $\theta$ always passes through all odd points of order 2: i.e., $\Omega\eta'+\eta'' \in \frac{1}{2}L_\Omega/L_\Omega$ such that $e_*(\eta) = -1$. One can count how many points of order 2 are even and how many odd, by induction on g. Thus they divide up like this:

| | | | even pts | | odd pts |
|---|---|---|---|---|---|
| g=1: | 4 | = | 3 | + | 1 |
| g=2: | 16 | = | 10 | + | 6 |
| g=3: | 64 | = | 36 | + | 28 |
| general genus: | $2^{2g}$ | = | $2^{g-1}(2^g+1)$ | + | $2^{g-1}(2^g-1)$ |

This is a nice exercise. For large g, there are nearly half of each kind.

Using the Theorem, we likewise divide up $\Sigma$ into even and odd parts: $\Sigma_+ \cup \Sigma_-$. The middle lines in (3.15) show that this division of $\Sigma$ does not depend on the choice of $A_i, B_i$ and all the rest: it depends merely on the multiplicity at 0 of the locus of points $I(P_1+\cdots+P_{g-1}-D)$ in Jax X. In particular, if $D \in \Sigma_-$, it follows that $D \equiv P_1+\cdots+P_{g-1}$ for some $P_i \in X$. In fact, $\Sigma_+$ and $\Sigma_-$ have a simple meaning in terms of the function theory on X, which we describe without proofs. This depends on a further theorem of Riemann which complements (3.6):

<u>Theorem</u> 3.16: <u>For all</u> $P_1,\cdots,P_{g-1} \in X$, <u>let</u> $\vec{e} = \vec{\Delta} - \sum_1^{g-1} \int_{P_0}^{P_i} \vec{\omega}$.
<u>Let</u> $\mathcal{L}(\Sigma P_i)$ <u>be the vector space of meromorphic functions on</u> X <u>with at most simple poles at</u> $P_1,\cdots,P_{g-1}$ <u>(or higher poles if several</u> $P_i$'s <u>coincide, the order of pole bounded by the multiplicity of</u> $P_i$). <u>Then</u>

$$\dim \mathcal{L}\Big(\sum_{i=1}^{g-1} P_i\Big) = \begin{pmatrix}\text{multiplicity of the zero of} \\ \vartheta(\vec{z}) \ \underline{\text{at}} \ \vec{z} = \vec{e}\end{pmatrix} .$$

Corollary 3.17: For all $D \in \Sigma$:

$$D \in \Sigma_+ \iff \left(\begin{array}{l}\text{either } \nexists\, P_1,\cdots,P_{g-1} \in X \text{ such that } D \equiv P_1+\cdots+P_{g-1} \\ \text{or if such } P_i \text{ exist, } \dim \mathcal{L}(P_1+\cdots+P_{g-1}) \text{ is even}\end{array}\right)$$

$$D \in \Sigma_- \iff \left(\begin{array}{l}\exists\, P_1,\cdots,P_{g-1} \in X \text{ such that } D \equiv P_1+\cdots+P_{g-1} \\ \text{and } \dim \mathcal{L}(P_1+\cdots+P_{g-1}) \text{ is odd}\end{array}\right)$$

For "almost all" X, it can be shown that in fact if $D \in \Sigma_+$, the $P_i$'s don't exist and if $D \in \Sigma_-$, $\mathcal{L}(P_1+\cdots+P_{g-1})$ contains only constants, hence is 1-dimensional.

Corollary 3.18: $D_o$ and hence $\vec{\Delta}$ can be determined from the function theory of X by the property:

$\forall\ E \in (\text{Pic } X)_2$, let $I(E) = \Omega\eta'+\eta''$, $\eta',\eta'' \in \frac{1}{2}\mathbb{Z}^g$. Then

(*) $$\dim \mathcal{L}(D_o+E) \equiv 4\,{}^t\eta'\cdot\eta''\ (\text{mod } 2).$$

Proof: In the bijections of 3.10, we have:

$$(D_o+E) \longleftrightarrow \text{zeroes of } \vartheta\left[\begin{smallmatrix}\eta'\\ \eta''\end{smallmatrix}\right](\vec{z})$$

so we have just expressed in (*) that even and odd elements should correspond. This characterizes $D_o$ because if $a',a'' \in \frac{1}{2}\mathbb{Z}^g$ and

$$4\,{}^t(\eta'+a')\cdot(\eta''+a'') \equiv 4\,{}^t\eta'\cdot\eta''\ (\text{mod } 2)$$

for all $\eta',\eta''$, then one sees that $a',a'' \in \mathbb{Z}^g$ in fact. QED

## §4. Siegel's Symplectic Geometry.

The other direction in which the theory of theta functions develops is the analysis of $\vartheta(\vec{z},\Omega)$, esp. $\vartheta(0,\Omega)$, as a function of $\Omega$. Before describing these results, however, we must understand the Siegel upper half space $\mathfrak{H}_g$ better. It is convenient to view $\mathfrak{H}_g$ in several ways, e.g., both as an explicit domain in $\mathbb{C}^{g(g+1)/2}$ and in a coordinate-free abstract way too. We base our analysis on a very useful elementary lemma in linear algebra:

Lemma 4.1: Let $A: \mathbb{R}^{2g} \times \mathbb{R}^{2g} \longrightarrow \mathbb{R}$ be the skew-symmetric form

$$A((x_1,x_2),(y_1,y_2)) = {}^t x_1 \cdot y_2 - {}^t x_2 \cdot y_1.$$

Then the following data on $\mathbb{R}^{2g}$ are equivalent:

a) a complex structure $J: \mathbb{R}^{2g} \longrightarrow \mathbb{R}^{2g}$ (i.e., a linear map with $J^2 = -I$) such that $A = \mathrm{Im}\, H$, $H$ a positive definite Hermitian form for this complex structure. (The existence of $H$ is equivalent to:

$$A(Jx,Jy) = A(x,y) \qquad \text{all } x,y \in \mathbb{R}^{2g}$$
$$A(Jx,x) > 0 \qquad \text{all } x \in \mathbb{R}^{2g}-(0)),$$

b) a homomorphism $i: \mathbb{Z}^{2g} \longrightarrow V$, $V$ a complex vector space, plus a positive definite Hermitian form on $V$ such that

$$\mathrm{Im}\, H(ix,iy) = A(x,y),$$

c) a $g$-dimensional complex subspace $P \subset \mathbb{C}^{2g}$ such that

$$A_{\mathbb{C}}(x,y) = 0, \text{ all } x,y \in P$$
$$iA_{\mathbb{C}}(x,\bar{x}) < 0, \text{ all } x \in P-(0)$$

($A_{\mathbb{C}}$ = complex linear extension of $A$),

d) *a* $g \times g$ *complex symmetric matrix* $\Omega$ *with* $\mathrm{Im}\,\Omega$ *positive definite*.

*The links between these data are:*

$a \longrightarrow a$ $H(x,y) = A(Jx,y) + iA(x,y)$

$a \longrightarrow b$ $i\colon\ \mathbb{Z}^{2g} \subset (\mathbb{R}^{2g}, J) = V$

$a \longrightarrow c$ $P =$ *locus of points* $ix - Jx$

$b \longrightarrow a$ $i$ *induces by* $\otimes\,\mathbb{R}$: $\mathbb{R}^{2g} \xrightarrow{\sim} V$, *hence a complex structure on* $\mathbb{R}^{2g}$

$b \longrightarrow c$ $i$ *induces by* $\otimes\,\mathbb{C}$: $\mathbb{C}^{2g} \twoheadrightarrow V$ *with kernel* $P$

$b \longrightarrow d$ *Coordinatize* $V$ *so that* $i((g+k)^{th}$ unit vector$) = k^{th}$ unit vector. *Then* $k^{th}$ column of $\Omega = i(k^{th}$ unit vector$)$.

$c \longrightarrow a$ *the complex structure on* $\mathbb{R}^{2g}$ *comes from*

$$\mathbb{R}^{2g} \subset \mathbb{C}^{2g} \twoheadrightarrow \mathbb{C}^{2g}/P$$

*and* $i$ *is*

$c \longrightarrow b$ $$\mathbb{Z}^{2g} \subset \mathbb{C}^{2g} \twoheadrightarrow \mathbb{C}^{2g}/P$$

$c \longrightarrow d$ $\Omega$ *is defined by* $(0,\cdots,\underset{i^{th}\text{ spot}}{1},\cdots,0;\ -\Omega_{i1},-\Omega_{i2},\cdots,-\Omega_{ig}) \in P$

$d \longrightarrow c$ $P =$ *set of points* $(x_1, -\Omega x_1)$

$d \longrightarrow a$ *the complex structure on* $\mathbb{R}^{2g}$ *is induced by requiring* $\underline{x} = \Omega x_1 + x_2$ *to be complex coordinates*

$$H((0,x_2),(0,y_2)) = {}^t x_2 \cdot (\mathrm{Im}\,\Omega)^{-1} \cdot y_2 \quad .$$

$d \longrightarrow b$ *We may set* $V = \mathbb{C}^g$

$$i(n_1,n_2) = \Omega n_1 + n_2 .$$

The reader is advised to study this lemma until the different facets of this structure are quite familiar. The proofs of the equivalences are easy.

It is clear from a, b or c that the symplectic group $Sp(2g,\mathbb{R})$ acts on this set of equivalent data. Thus if $\gamma \in Sp(2g,\mathbb{R})$, then $\gamma$ carries

$$J \longmapsto J' = \gamma J \gamma^{-1}$$

$$(V,i) \longmapsto (V,i') \quad \text{where} \quad i' = i \circ \gamma^{-1}$$

$$P \longmapsto P' = \gamma(P).$$

Now we can write such a $\gamma$ as:

$$\gamma = \begin{pmatrix} A & B \\ C & D \end{pmatrix}, \qquad A,B,C,D \quad n \times n \quad \text{real matrices}$$

such that:

$${}^t\begin{pmatrix} A & B \\ C & D \end{pmatrix}\begin{pmatrix} 0 & I \\ -I & 0 \end{pmatrix}\begin{pmatrix} A & B \\ C & D \end{pmatrix} = \begin{pmatrix} 0 & I \\ -I & 0 \end{pmatrix}.$$

Therefore

$$P' = \gamma(P) = \left\{ \begin{pmatrix} A & B \\ C & D \end{pmatrix}\begin{pmatrix} x_1 \\ -\Omega x_1 \end{pmatrix} \middle| \ x_1 \in \mathbb{C}^g \right\}$$

$$= \left\{ \begin{pmatrix} Ax_1 - B\Omega x_1 \\ Cx_1 - D\Omega x_1 \end{pmatrix} \middle| \ x_1 \in \mathbb{C}^g \right\}.$$

But this is the space of points $\begin{pmatrix} y_1 \\ -\Omega' y_1 \end{pmatrix}$, $y_1 \in \mathbb{C}^g$, so

$$\Omega' = (D\Omega - C)(-B\Omega + A)^{-1}.$$

This defines therefore an action of $Sp(2g,\mathbb{R})$ on $\mathfrak{H}_g$. Note that every symplectic matrix $\begin{pmatrix} A & B \\ C & D \end{pmatrix}$ acts by a bi-holomorphic map $\mathfrak{H}_g \longrightarrow \mathfrak{H}_g$.

We may make the formula look more familiar if we compose with the automorphism $\gamma \longmapsto {}^t\gamma^{-1}$ of $Sp(2,\mathbb{R})$. In fact

$$ {}^t\gamma \begin{pmatrix} 0 & I \\ -I & 0 \end{pmatrix} \gamma = \begin{pmatrix} 0 & I \\ -I & 0 \end{pmatrix} $$

so

$$ \begin{aligned} {}^t\gamma^{-1} &= \begin{pmatrix} 0 & I \\ -I & 0 \end{pmatrix} \gamma \begin{pmatrix} 0 & -I \\ I & 0 \end{pmatrix} \\ &= \begin{pmatrix} D & -C \\ -B & A \end{pmatrix} \end{aligned} $$

Thus after composing with this automorphism of $Sp(2g,\mathbb{R})$, the action is:

$$ \Omega \longmapsto \Omega' = (A\Omega+B)(C\Omega+D)^{-1}. $$

In particular, this shows that $(A\Omega+B)(C\Omega+D)^{-1}$ is symmetric. Another way of phrasing the result is this:

Lemma 4.2: *Let* $i: \mathbb{Z}^{2g} \longrightarrow V$ *be data as in* (b) *above corresponding to* $\Omega$ *in* (d). *Let* $\vec{e}_1,\cdots,\vec{e}_{2g}$ *be the unit vectors in* $\mathbb{Z}^{2g}$ *and let*

$$ \begin{aligned} \vec{e}_i{}' &= \sum A_{ij}\vec{e}_j + \sum B_{ij}\vec{e}_{j+g} \\ \vec{e}_{i+g}{}' &= \sum C_{ij}\vec{e}_j + \sum D_{ij}\vec{e}_{j+g} \end{aligned} $$

*be a new symplectic basis of* $\mathbb{Z}^{2g}$. *Let* $i'$ *be the composition*

$$\begin{array}{ccccc} \mathbb{Z}^{2g} & \longrightarrow & \mathbb{Z}^{2g} & \xrightarrow{i} & V \\ \vec{e}_i & \longmapsto & \vec{e}_i' & . & \end{array}$$

<u>Then</u> $i': \mathbb{Z}^{2g} \longrightarrow V$ <u>corresponds to</u> $(A\Omega+B)(C\Omega+D)^{-1}$.

We reprove this to get another handle on the situation: we may let $V = \mathbb{C}^g$ and let $i$ be

$$i(\vec{n}_1,\vec{n}_2) = \Omega\vec{n}_1+\vec{n}_2.$$

Then

$$\begin{aligned} i(\vec{e}_i')_k &= \sum_j A_{ij}\Omega_{kj} + B_{ik} \\ &= (A\Omega+B)_{ik} \end{aligned}$$

by the symmetry of $\Omega$. Likewise,

$$i(\vec{e}_{i+g}')_k = (C\Omega+D)_{ik}.$$

Therefore

$$i'(\vec{n}_1,\vec{n}_2)_k = \sum (n_1)_i (A\Omega+B)_{ik} + (n_2)_i (C\Omega+D)_{ik}.$$

We change the identification of $V$ with $\mathbb{C}^g$ by composing with the automorphism ${}^t(C\Omega+D)^{-1}$:

$$(z_1,\cdots,z_g) \longmapsto \Big( \sum_k (C\Omega+D)^{-1}_{k1} z_k, \cdots, \sum_k (C\Omega+D)^{-1}_{kg} z_k \Big).$$

Composing, we find

$$i'(\vec{n}_1,\vec{n}_2)_k = \sum (n_1)_i [(A\Omega+B)(C\Omega+D)^{-1}]_{ik} + (n_2)_k .$$

Because of the symmetry of $(A\Omega+B)(C\Omega+D)^{-1}$, proven above, this can be written

$$i'(\vec{n}_1,\vec{n}_2) = (A\Omega+B)(C\Omega+D)^{-1}\cdot\vec{n}_1 + \vec{n}_2. \qquad \underline{\text{QED}}$$

Here is a diagram that summarizes in terms of matrices the situation relating i and i' constructed via $\Omega,\Omega'$:

$$\begin{array}{ccc}
(x_1,x_2) & \longmapsto & \Omega x_1+x_2 \\
\mathbb{Z}^{2g} & \xrightarrow{\quad i \quad} & \mathbb{C}^g \\
{}^t\begin{pmatrix} A & B \\ C & D \end{pmatrix} \Big\uparrow \Big\downarrow \begin{pmatrix} D & -C \\ -B & A \end{pmatrix} & & {}^t(C\Omega+D) \Big\uparrow \Big\downarrow {}^t(C\Omega+D)^{-1} = -\Omega' C+A \\
\mathbb{Z}^{2g} & \xrightarrow{\quad i' \quad} & \mathbb{C}^g \\
(x_1,x_2) & \longmapsto & \Omega' x_1+x_2
\end{array} \tag{4.3}$$

$$\Omega' = (A\Omega+B)(C\Omega+D)^{-1}, \quad \begin{pmatrix} A & B \\ C & D \end{pmatrix} \in Sp(2g,\mathbb{Z}) .$$

(We verify

a) that ${}^t\begin{pmatrix} A & B \\ C & D \end{pmatrix}, \begin{pmatrix} D & -C \\ -B & A \end{pmatrix}$ are inverse to each other because $\begin{pmatrix} A & B \\ C & D \end{pmatrix}$ being symplectic means

$$(*) \qquad {}^t\begin{pmatrix} A & B \\ C & D \end{pmatrix} \begin{pmatrix} 0 & I \\ -I & 0 \end{pmatrix} \begin{pmatrix} A & B \\ C & D \end{pmatrix} = \begin{pmatrix} 0 & I \\ -I & 0 \end{pmatrix}$$

b) that $^t(C\Omega+D)^{-1} = -\Omega' C+A$ because $^t\Omega' = \Omega'$ so this means

$$^t(C\Omega+D)^{-1} = -^t(C\Omega+D)^{-1}\cdot {}^t(A\Omega+B)\cdot C + A$$

which reduces to

$$^tAC = {}^tCA \quad \text{and} \quad {}^tDA - {}^tBC = I_g$$

which in turn follow immediately from (*).

The commutativity is straightforward. )

We note for later use the following consequence:

Proposition 4.4: The group $Sp(2g,\mathbb{R})$ acts on the space $\mathbb{C}^g \times \mathfrak{H}_g$ by the maps:

$$(\vec{z},\Omega) \longmapsto ({}^t(C\Omega+D)^{-1}\cdot\vec{z},\ (A\Omega+B)(C\Omega+D)^{-1}).$$

Proof: For elements of $Sp(2g,\mathbb{Z})$, the fact that this is a group action follows from diagram (4.3), by putting together the diagrams for i and i', i' and i". In fact, the same diagram holds if $\mathbb{Z}$ is replaced by $\mathbb{R}$, by the same argument, so we need not restrict to elements of $Sp(2g,\mathbb{Z})$. QED

We can use the action of $Sp(2g,\mathbb{R})$ on $\mathfrak{H}_g$ to give a purely group-theoretic definition of $\mathfrak{H}_g$ as a coset space. In fact:

Proposition 4.5. $Sp(2g,\mathbb{R})$ acts transitively on $\mathfrak{H}_g$ and the stabilizer of $iI_g$ is isomorphic to $U(g,\mathbb{C})$, embedded in $Sp(2g,\mathbb{R})$ by

$$X \longmapsto \begin{pmatrix} \mathrm{Re}\, X & \mathrm{Im}\, X \\ -\mathrm{Im}\, X & \mathrm{Re}\, X \end{pmatrix} .$$

Thus $\mathfrak{H}_g \cong Sp(2g,\mathbb{R})/U(g,\mathbb{C})$.

Proof: The transitivity is checked quickly by introducing 2 elementary subgroups in $Sp(2g,\mathbb{R})$:

(I) $\quad \gamma = \begin{pmatrix} A & 0 \\ 0 & {}^tA^{-1} \end{pmatrix}, \quad A \in GL(g,\mathbb{R})$.

(II) $\quad \gamma = \begin{pmatrix} I & B \\ 0 & I \end{pmatrix}, \quad$ B any $g \times g$ real symmetric matrix.

In fact, $\gamma$ as in I acts by $(x_1,x_2) \longmapsto (Ax_1, {}^tA^{-1}x_2)$, and

$$\begin{aligned} A((\gamma x_1,\gamma x_2),(\gamma y_1,\gamma y_2)) &= {}^t(\gamma x_1)\cdot(\gamma y_2) - {}^t(\gamma x_2)\cdot(\gamma y_1) \\ &= {}^t(Ax_1)\cdot({}^tA^{-1}y_2) - {}^t({}^tA^{-1}x_2)\cdot(Ay_1) \\ &= {}^tx_1\cdot y_2 - {}^tx_2\cdot y_1 \\ &= A((x_1,x_2),(y_1,y_2)), \end{aligned}$$

while $\gamma$ as in II acts by $(x_1,x_2) \longmapsto (x_1+Bx_2,x_2)$, and

$$\begin{aligned} A((\gamma x_1,\gamma x_2),(\gamma y_1,\gamma y_2)) &= {}^t(x_1+Bx_2)\cdot y_2 - {}^tx_2\cdot(y_1+By_2) \\ &= {}^tx_1\cdot y_2 + {}^tx_2{}^tBy_2 - {}^tx_2\cdot y_1 - {}^tx_2By_2 \\ &= A((x_1,x_2),(y_1,y_2)). \end{aligned}$$

Acting on $\mathfrak{H}_g$, these maps carry

$$\Omega \longmapsto A\Omega\cdot{}^tA$$

and

$$\Omega \longmapsto \Omega+B.$$

Hence together they carry $iI_g$ to $i(A\cdot {}^tA) + B$. Since any positive symmetric matrix can be written $A\cdot {}^tA$, this gives any element of $\mathfrak{h}_g$. This proves transitivity.

The stabilizer of a point of $\mathfrak{h}_g$ is most easily identified in version (a), lemma 4.1. Here $(\mathbb{R}^{2g},J)$ is a complex vector space and $A = \mathrm{Im}\, H$, $H$ positive definite Hermitian. The symplectic automorphisms of $\mathbb{R}^{2g}$ that also commute with J are the complex-linear automorphisms of $(\mathbb{R}^{2g},J)$. Since $A = \mathrm{Im}\, H$, these must preserve H. So this is the group of unitary automorphisms of $(\mathbb{R}^{2g},J,H)$. In particular, if $\Omega = iI_g$, then $z_k = i(x_1)_k + (x_2)_k$, $1 \le k \le g$, are complex coordinates on $\mathbb{R}^{2g}$, so:

$$J = \begin{pmatrix} 0 & +I \\ -I & 0 \end{pmatrix}.$$

Then

$$\begin{pmatrix} A & B \\ C & D \end{pmatrix} J = J \begin{pmatrix} A & B \\ C & D \end{pmatrix}$$

if and only if $A = D$, $B = -C$ and as

$$\begin{pmatrix} A & B \\ -B & A \end{pmatrix} \begin{pmatrix} n_1 \\ n_2 \end{pmatrix} = \begin{pmatrix} An_1+Bn_2 \\ -Bn_1+An_2 \end{pmatrix},$$

it carries $in_1+n_2$ to $i(An_1+Bn_2) + (-Bn_1+An_2) = (A+iB)(in_1+n_2)$. This proves that the stabilizer of $iI_g$ is as claimed. QFD

One can now proceed to build up a detailed "symplectic geometry" on $\mathfrak{h}_g$: first one defines a metric on $\mathfrak{h}_g$ which is invariant by the action of $Sp(2g,\mathbb{R})$. In this metric, one can describe geodesics, compute curvature, investigate totally geodesic subspaces, etc. A good reference is

Siegel, Symplectic Geometry, Academic Press, 1964.
This is a generalization of the non-Euclidean metric on the upper-half plane H, which is $SL(2,\mathbb{R})$-invariant, hence has constant curvature. Our interest however is in the action of the subgroup $Sp(2g,\mathbb{Z})$ of $Sp(2g,\mathbb{R})$ on $\mathfrak{H}_g$. Because

a) $Sp(2g,\mathbb{Z}) \subset Sp(2g,\mathbb{R})$ is discrete, i.e., $\exists$ a neighborhood

$$U = \{X \mid |X_{ij}-\delta_{ij}| < 1\}$$

of the identity meeting $Sp(2g,\mathbb{Z})$ only in $I_{2g}$,

and

b) The stabilizer of a point of $\mathfrak{H}_g$ is a compact Lie group,

it follows:

c) $Sp(2g,\mathbb{Z})$ acts discontinuously on $\mathfrak{H}_g$: (1) $\forall\ x \in \mathfrak{H}_g$, $S_x = \{\gamma \in Sp(2g,\mathbb{Z}) \mid \gamma x = x\}$ is finite and $\exists$ a neighborhood $U_x$ of $x$ which is stable under $S_x$ such that $\forall\ \gamma \in Sp(2g,\mathbb{Z})$:

$$\gamma U_x \cap U_x \neq \emptyset \iff \gamma \in S_x .$$

and (2), $\forall\ x,y \in \mathfrak{H}_g$ such that $x \neq \gamma y$ for any $\gamma \in Sp(2g,\mathbb{Z})$, $\exists$ neighborhoods $U_x, U_y$ of $x,y$ such that $U_x \cap \gamma U_y = \emptyset$, all $\gamma \in Sp(2g,\mathbb{Z})$.

Proof of c : Because $U(g,\mathbb{C})$ is compact,

$$\pi: \ Sp(2g,\mathbb{R}) \longrightarrow Sp(2g,\mathbb{R})/U(g,\mathbb{C}) \xrightarrow{\sim} \mathfrak{H}_g$$

is proper, i.e., the inverse image of a compact set is compact (in fact, in this case, we get a section by considering $\gamma$'s of form $\begin{pmatrix} I & B \\ 0 & I \end{pmatrix}\begin{pmatrix} A & 0 \\ 0 & A^{-1} \end{pmatrix}$, $A$ positive definite symmetric -see proof of 4.5- hence as a topological space

$$Sp(2g,\mathbb{R}) \cong U(g,\mathbb{C}) \times [Sp(2g,\mathbb{R})/U(g,\mathbb{C})].)$$

Let $U_x^{(1)}$ be a relatively compact neighborhood of $x$: then $\pi^{-1}(\overline{U}_x^{(1)})$ is compact, hence

$$W = \{\gamma_1 \cdot \gamma_2^{-1} \mid \gamma_1, \gamma_2 \in \pi^{-1}(\overline{U}_x^{(1)})\}$$

is compact. Hence the intersection $W \cap Sp(2g,\mathbb{Z})$ is compact and discrete, hence is finite. But this is the same as the set

$$F = \{\gamma \in Sp(2g,\mathbb{Z}) \mid \gamma(\overline{U}_x^{(1)}) \cap \overline{U}_x^{(1)} \neq \emptyset\} .$$

Now $F = S_x \cup F_1$, where if $\gamma \in F_1$, then $\gamma x \neq x$. For all $\gamma \in F_1$, let $U_\gamma, V_\gamma$ be disjoint open neighborhoods of $x$, $\gamma x$. Let

$$U_x^{(2)} = \Big(\bigcap_{\gamma \in F_1} (U_\gamma \cap \gamma^{-1} V_\gamma)\Big) \cap U_x^{(1)} .$$

One checks that $\gamma U_x^{(2)} \cap U_x^{(2)} \neq \emptyset$ only if $\gamma x = x$. The sought-for $U_x$ can be taken to be

$$U_x^{(3)} = \bigcap_{\gamma \in S_x} \gamma U_x^{(2)} .$$

This proves (1). We leave the proof of (2) to the reader. QED

One then considers the orbit space

$$\mathcal{A}_g = \mathfrak{h}_g / Sp(2g,\mathbb{Z}) .$$

By definition, a subset of $\mathcal{A}_g$ is open if its inverse image in $\mathfrak{h}_g$ is open. By (c), each of the local quotients $U_x/S_x$ is an open subset of $\mathcal{A}_g$ and the induced topology is immediately seen to be the quotient topology. Each of these local pieces is an open set in $\mathbb{C}^{g(g+1)/2}$ modulo a finite group of analytic automorphisms, hence it is an analytic space. Together these give an analytic structure on $\mathcal{A}_g$. $\mathcal{A}_g$ is called the <u>Siegel modular variety</u> — a very special but most interesting space. (c2) tells us that $\mathcal{A}_g$ is a Hausdorff space.

We conclude this section by tying $\mathcal{A}_g$ together with the theory of projectively embeddable complex tori, generalizing the ideas of Ch. I, §12. We need some preliminaries on cohomology. For our purposes, we will use DeRham cohomology: on any oriented compact manifold M,

$$H^k(M,\mathbb{R}) \cong \frac{\text{space of closed exterior k-forms}}{\text{space of exact exterior k-forms}}$$

and in this isomorphism, the following subspaces correspond:

$$H^k(M,\mathbb{Z})/\text{torsion} \cong \frac{\left(\begin{matrix}\text{space of closed k-forms } \omega \text{ with}\\ \text{integral periods around all k-cycles}\end{matrix}\right)}{\text{exact forms}}$$

In particular, if V is a real vector space and $L \subset V$ is a lattice, then for any k elements $\lambda_1,\cdots,\lambda_k \in L$, we get a k-cycle $\sigma(\lambda_1,\cdots,\lambda_k)$ on V/L consisting of the image of the k-dimensional cube:

$$\left\{ \sum_{i=1}^{k} t_i \lambda_i \in V \;\middle|\; 0 \le t_i \le 1, \text{ for all } i \right\} \longrightarrow V/L.$$

Then taking periods on $\sigma(\lambda_1, \cdots, \lambda_k)$, we get an isomorphism

$$H^k(V/L, \mathbb{Z}) \xrightarrow{\approx} \left\{ \begin{array}{c} \text{multi-linear alternating forms} \\ A: \overbrace{L \times \cdots \times L}^{k \times} \longrightarrow \mathbb{Z} \end{array} \right\}$$

On the other hand, for complex projective space $\mathbb{P}^n$:

$$H^k(\mathbb{P}^n, \mathbb{Z}) = \begin{cases} 0 & \text{if } k \text{ odd} \quad \text{or} \quad k > 2n \\ \mathbb{Z} & \text{if } k \text{ even, } 0 \le k \le 2n \end{cases}$$

where if k is even the identification is given by integrating over linear subspace $L \subset \mathbb{P}^n$, $\dim L = k/2$. For our purposes, we need only to have an expression for a 2-form $\omega$ on $\mathbb{P}^n$ representing the generator of $H^2(\mathbb{P}^n, \mathbb{Z})$. The simplest is:

$$\omega = \frac{i}{2\pi} \sum_{i,j=0}^{n} \frac{\partial^2}{\partial z_i \partial \bar{z}_j} \log \left( \sum_{i=0}^{n} |z_i|^2 \right) dz_i \wedge d\bar{z}_j$$

(where $z_0, \cdots, z_n$ are homogeneous coordinates: i.e., coordinates in $\mathbb{C}^{n+1}$ with the canonical map

$$\pi: \mathbb{C}^{n+1} - (0) \longrightarrow \mathbb{P}^n$$

and what we have written here is $\pi^* \omega$).

Now if $\Omega \in \mathfrak{h}_g$, let

$$i: \mathbb{Z}^{2g} \longrightarrow V$$

be data (b) associated to $\Omega$ as in Lemma 4.1. Explicitly, we may set $V = \mathbb{C}^g$, $i(n_1,n_2) = \Omega n_1+n_2$. Then Image (i) is the lattice $L_\Omega$, and

$$V/i(\mathbb{Z}^{2g}) = X_\Omega .$$

Moreover, the alternating form $A: \mathbb{Z}^{2g} \times \mathbb{Z}^{2g} \longrightarrow \mathbb{Z}$ gives us, as we have just explained, a class

$$[A] \in H^2(X_\Omega,\mathbb{Z}) .$$

Note that if an element $\gamma \in Sp(2g,\mathbb{Z})$ acts on $\Omega$, then $i(\mathbb{Z}^{2g})$ is unchanged and the alternating form on $i(\mathbb{Z}^{2g})$ is unchanged. This gives us a well-defined map:

$$\mu: \mathcal{A}_g \longrightarrow \left\{ \begin{array}{l} \text{set of pairs } (X,[A]), \\ X \text{ a complex torus,} \\ [A] \in H^2(X,\mathbb{Z}) \end{array} \right\} \Big/ \left\{ \begin{array}{c} \text{modulo holomorphic} \\ \text{isomorphisms pre-} \\ \text{serving the cohomology} \\ \text{classes} \end{array} \right\}$$

We can easily prove that $\mu$ is injective:

let

$$\phi: X_{\Omega_1} \xrightarrow{\sim} X_{\Omega_2}$$

be an isomorphism such that $\phi^*([A_2]) = [A_1]$. Write $X_{\Omega_i} = \mathbb{C}^g/L_{\Omega_i}$ and lift $\phi$ to an isomorphism of universal covering spaces, and of fundamental groups:

$$\tilde{\phi}: \mathbb{C}^g \xrightarrow{\sim} \mathbb{C}^g,$$

$$\phi_*: L_{\Omega_1} \xrightarrow{\sim} L_{\Omega_2}$$

where

$$(*) \qquad \tilde{\phi}(z+a) = \tilde{\phi}(z) + \phi_*(a), \quad \text{all} \quad a \in L_{\Omega_1}.$$

Then for all i,

$$\frac{\partial\tilde{\phi}}{\partial z_i}(z+a) = \frac{\partial\tilde{\phi}}{\partial z_i}(z) \quad .$$

Hence $\frac{\partial\tilde{\phi}}{\partial z_i}$ is a holomorphic function on $\mathbb{C}^g$, periodic with respect to $L_{\Omega_1}$, hence bounded. Then $\frac{\partial\tilde{\phi}}{\partial z_i}$ must be a constant, so $\tilde{\phi}$ is linear plus a constant. We may throw out the constant without affecting (*). Note too that $\phi^*([A_2]) = [A_1]$ implies

$$A_2(\phi_*x, \phi_*y) = A_1(x,y).$$

Thus we have isomorphisms:

$$\begin{array}{ccc} \mathbb{Z}^{2g} & \xrightarrow{i_1} & \mathbb{C}^g \\ \phi_* \downarrow & & \downarrow \tilde{\phi} \\ \mathbb{Z}^{2g} & \xrightarrow{i_2} & \mathbb{C}^g \end{array}$$

where $\phi_*$ is symplectic. This means exactly that an element of $Sp(2g,\mathbb{Z})$ carries $(\mathbb{C}^g, i)$ to $(\mathbb{C}^g, i')$, hence $\Omega$ to $\Omega'$. Thus $\mu$ is injective. To describe the image, we make the following definition:

Definition 4.6. Let $X$ be a complex torus, $[A] \in H^2(X,\mathbb{Z})$. Then $[A]$ is a principal polarization of $X$ if

a) Expressed as a skew-symmetric integral matrix, $\det[A] = 1$,

b) there is a holomorphic embedding

$$f: X \hookrightarrow \mathbb{P}^n$$

such that if $[\omega] \in H^2(\mathbb{P}^n,\mathbb{Z})$ is the positive generator, then

$$f^*[\omega] = N.[A],$$

some $N \geq 1$.

Theorem 4.7. The map $\mu$ is bijective between $\mathfrak{A}_g$ and the set of isomorphism classes of principally polarized complex tori $(X,[A])$.

Having described the projective embeddings of $X_\Omega$ in §1, we can easily show that $[A]$ is a principal polarization on $X_\Omega$. We shall omit the proof that all principally polarized tori are isomorphic to an $X_\Omega$: it is a variant of Theorem 1.3, part (3) and, like that result, is proven in the author's book Abelian Varieties, §§2 and 3.

To prove $[A]$ is a principal polarization, we embed $X_\Omega$ in $\mathbb{P}^{n^{2g}-1}$ by the method of §1:

$$\vec{z} \longmapsto \left(\cdots\cdots, \vartheta\begin{bmatrix} a_i \\ b_i \end{bmatrix}(n\vec{z},\Omega), \cdots\cdots\right)$$

where $[{a_i \atop b_i}]$ run over cosets of $\mathbb{Z}^{2g}$ in $\frac{1}{n}\mathbb{Z}^{2g}$ and n a fixed integer, $n \geq 2$. (This is the embedding of §1 for $\mathbb{C}^g/nL_\Omega$ adapted to $\mathbb{C}^g/L_\Omega$ by the remark at the end of §1.) We have a diagram of maps:

$$\begin{array}{ccc} \mathbb{C}^g & \longrightarrow & \mathbb{C}^{n^{2g}}-(0) \\ \pi\downarrow & & \downarrow \\ X_\Omega & \xhookrightarrow[f]{} & \mathbb{P}^{n^{2g}-1} \end{array}$$

Thus $\pi^*f^*[\omega]$ is represented by the 2-form on $\mathbb{C}^g$

$$\eta = \frac{i}{2\pi} \sum \frac{\partial^2}{\partial z_i \partial \bar{z}_j} \log\left(\sum_i |\vartheta[{a_i \atop b_i}](n\vec{z},\Omega)|^2\right) dz_i \wedge d\bar{z}_j \ .$$

We need only compute the periods of $\eta$ to complete the proof.

<u>Lemma</u> 4.8. $\displaystyle\int_{\sigma(\Omega n_1+n_2,\Omega m_1+m_2)} \eta = n^2\cdot({}^tn_1\cdot m_2 - {}^tn_2\cdot m_1)$.

<u>Proof</u>: The functional equation for $\vartheta[{a_i \atop b_i}]$ shows that

$$\left|\vartheta[{a_i \atop b_i}](n(\vec{z}+\Omega n_1+n_2))\right|^2 = \left(e^{-2\pi {}^tn_1\cdot \mathrm{Im}\Omega n_1 - 4\pi {}^tn_1(\mathrm{Im}\ z)}\right)^{n^2} \left|\vartheta[{a_i \atop b_i}](n\vec{z})\right|^2$$

hence

$$(*) \quad \log \sum_i \left|\vartheta[{a_i \atop b_i}](n(\vec{z}+\Omega n_1+n_2))\right|^2 = 2\pi n^2({}^tn_1\cdot \mathrm{Im}\ \Omega\cdot n_1 + 2\,{}^tn_1\cdot \mathrm{Im}\ z) + \log \sum_i \left|\vartheta[{a_i \atop b_i}](n\vec{z})\right|^2$$

We set

$$\zeta = \frac{i}{2\pi} \sum \frac{\partial}{\partial z_i} \log\Big(\sum_i \Big|\vartheta[{}^{a_i}_{b_i}](n\vec{z})\Big|^2\Big) dz_i \quad .$$

Then $d\zeta = -\eta$ (the $\frac{\partial}{\partial z_i}$ terms cancel out in $d\zeta$), and writing $\operatorname{Im} z = (z-\bar{z})/2i$, we find:

$$\frac{\partial}{\partial z_i}(\operatorname{Im} z) = \frac{1}{2i}$$

hence differentiating (*):

$$\zeta(\vec{z} + \Omega n_1 + n_2) = -n^2({}^t n_1 \cdot dz) + \zeta(\vec{z}).$$

Now the rectangle $\sigma$ is

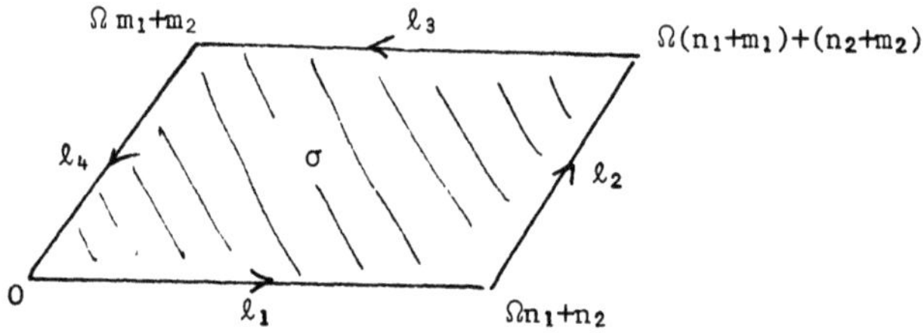


so by Green's theorem:

$$\int_\sigma \eta = \int_{\partial\sigma} (-\zeta) = \int_{\ell_1+\ell_3} (-\zeta) + \int_{\ell_2+\ell_4} (-\zeta)$$

$$= n^2 \int_{\ell_3} {}^t m_1 \cdot dz + n^2 \int_{\ell_2} {}^t n_1 \cdot dz$$

$$= n^2[{}^t m_1 \cdot (-\Omega n_1 - n_2) + {}^t n_1 \cdot (\Omega m_1 + m_2)]$$

$$= n^2({}^t n_1 \cdot m_2 - {}^t n_2 \cdot m_1).$$ QED

## §5. $\vartheta$ as a modular form.

We want to consider now the dependence of the function $\vartheta(\vec{z},\Omega)$ on $\Omega$. As in the one-variable theory, the fundamental fact is a functional equation for $\vartheta$ for the action of $Sp(2g,\mathbb{Z})$ on both the variables $\vec{z}$ and $\Omega$. As before, there is a rather tricky $8^{th}$ root of 1 in this equation. Without working this out, we can state the functional equation as:

$$\text{(5.1)}\qquad \begin{aligned} &\vartheta\left({}^t(C\Omega+D)^{-1}\cdot\vec{z},\,(A\Omega+B)(C\Omega+D)^{-1}\right) \\ &\quad = \zeta_\gamma\cdot\det(C\Omega+D)^{1/2}\cdot\exp[\pi i\,{}^t\vec{z}\cdot(C\Omega+D)^{-1}C\cdot\vec{z}]\cdot\vartheta(\vec{z},\Omega)\end{aligned}$$

where $\zeta_\gamma^8 = 1$, and

$$\gamma = \begin{pmatrix} A & B \\ C & D\end{pmatrix} \in \quad Sp(2g,\mathbb{Z})$$

satisfies

$$\text{diagonal } ({}^tAC) \text{ even}$$
$$\text{diagonal } ({}^tBD) \text{ even.}$$

This set of elements of $Sp(2g,\mathbb{Z})$ may be described as those $\gamma$ such that, modulo 2, $\gamma$ preserves the orthogonal form

$$Q(n_1,n_2) = {}^tn_1.n_2 \in (\mathbb{Z}/2\mathbb{Z})$$

as well as the alternating form A: see the Appendix. In particular this is a group, which we call $\Gamma_{1,2}$ following Igusa.

In the appendix, a set of generators of $\Gamma_{1,2}$ is found and using these, we may prove (5.1) in 4 steps:

a) Showing that if (5.1) holds for $\gamma_1, \gamma_2 \in Sp(2g,\mathbb{Z})$, then it holds for $\gamma_1\gamma_2$.

b) verifying (5.1) for $\begin{pmatrix} A & 0 \\ 0 & {}^tA^{-1} \end{pmatrix}$, $A \in GL(g,\mathbb{Z})$

c) verifying (5.1) for $\begin{pmatrix} I & B \\ 0 & I \end{pmatrix}$, B symmetric, even diagonal,

d) verifying (5.1) for $\begin{pmatrix} 0 & -I \\ I & 0 \end{pmatrix}$.

We may check (a) by a direct matrix computation, but perhaps a more interesting way is to reformulate (5.1) in terms of a closely related function $\vartheta^\alpha$ which is then to be shown to be $Sp(2g,\mathbb{Z})$-<u>invariant</u>. Since invariance by $\gamma_1$ and $\gamma_2$ obviously implies invariance by $\gamma_1\gamma_2$, (a) becomes obvious. The advantage of this approach is that the new function $\vartheta^\alpha$ must have a certain importance due to its invariance: this will be explored in Chapter IV.

To prove (5.1) for all $\vec{z}$, it certainly suffices to do so for $\vec{z} = \Omega\vec{n}_1+\vec{n}_2$, $\vec{n}_i \in \mathbb{Q}^g$. So as a first step, we substitute $\Omega\vec{n}_1+\vec{n}_2$ for $\vec{z}$ and rewrite (5.1) for $\vartheta\left[\begin{smallmatrix} n_1 \\ n_2 \end{smallmatrix}\right](0,\Omega)$. We claim that (5.1) is equivalent to:

$$\vartheta\begin{bmatrix} Dn_1-Cn_2 \\ -Bn_1+An_2 \end{bmatrix}(0,(A\Omega+B)(C\Omega+D)^{-1})$$

(5.2)

$$= \zeta_\gamma \cdot \det(C\Omega+D)^{1/2} \cdot \exp\left(-\pi i\,{}^tn_1 \cdot {}^tBD \cdot n_1 + 2\pi i\,{}^tn_1\,{}^tBCn_2 - \pi i\,{}^tn_2\,{}^tACn_2\right)\vartheta\left[\begin{smallmatrix} n_1 \\ n_2 \end{smallmatrix}\right](0,\Omega)$$

($\zeta_\gamma$ depending only on $\begin{pmatrix} A & B \\ C & D \end{pmatrix}$, not on $n_1, n_2, \Omega$.)

This calculation goes like this: you substitute $\Omega n_1+n_2$ for z, use the fact that

$$^t(C\Omega+D)^{-1}(\Omega n_1+n_2) = \Omega'(Dn_1-Cn_2) + (-Bn_1+An_2)$$

by (4.3), and then use the definition of $\vartheta[{n_1 \atop n_2}](0,\Omega)$. (5.2) follows except that one has a messy exponential factor, viz. exp of

$$\begin{aligned} &\pi i({}^tn_1\Omega+{}^tn_2)(C\Omega+D)^{-1}C(\Omega n_1+n_2) \\ &- \pi i\,{}^tn_1\Omega n_1 - 2\pi i\,{}^tn_1 n_2 \\ &+ \pi i({}^tn_1{}^tD-{}^tn_2{}^tC)(A\Omega+B)(C\Omega+D)^{-1}(Dn_1-Cn_2) \\ &+ 2\pi i({}^tn_1{}^tD-{}^tn_2{}^tC)\cdot(-Bn_1+An_2). \end{aligned}$$

In this, you separate the 4 terms $\pi i\,{}^tn_1\cdot(\ \ )n_1$, $\pi i\,{}^tn_1\cdot(\ \ )n_2$, $\pi i\,{}^tn_2\cdot(\ \ )n_1$, $\pi i\,{}^tn_2\cdot(\ \ )n_2$ and simplify each one, using the basic facts on $\begin{pmatrix} A & B \\ C & D\end{pmatrix}$ expressing that it is symplectic:

$$\begin{aligned} &{}^tD\cdot A - {}^tB\cdot C = I_g \\ &{}^tD\cdot B = {}^tB\cdot D,\ {}^tC\cdot A = {}^tA\cdot C \quad . \end{aligned}$$

For example, take the first. It is

$$\begin{aligned} &\pi i\,{}^tn_1\{\ \Omega(C\Omega+D)^{-1}((C\Omega+D)-D) - \Omega + {}^tD(A\Omega+B)(C\Omega+D)^{-1}D - 2\,{}^tDB\}n_1 \\ = \ &\pi i\,{}^tn_1\{\Omega-\Omega(C\Omega+D)^{-1}\cdot D - \Omega + {}^tD(A\Omega+B)(C\Omega+D)^{-1}D - 2\,{}^tDB\}n_1 \\ = \ &\pi i\,{}^tn_1\{(-\ \Omega + {}^tDA\Omega + {}^tDB)(C\Omega+D)^{-1}D - 2\,{}^tDB\}n_1 \\ = \ &\pi i\,{}^tn_1\{({}^tB\cdot C\Omega + {}^tBD)(C\Omega+D)^{-1}D - 2\,{}^tBD\}n_1 \\ = \ &\pi i\,{}^tn_1\{-{}^tBD\}\cdot n_1 \ . \end{aligned}$$

The others are similar.

However, if we examine the above calculation we see that it would have come out even simpler if, instead of

$$\vartheta[{}^{n_1}_{n_2}](0,\Omega) = \exp[\pi i\, {}^t n_1 \Omega n_1 + 2\pi i\, {}^t n_1 n_2]\, \vartheta(\Omega n_1 + n_2, \Omega)$$

we use a modified $\vartheta$ that we will call $\vartheta^\alpha$:

$$(5.3) \qquad \vartheta^\alpha[{}^{n_1}_{n_2}](\Omega) = \exp[\pi i\, {}^t n_1 \Omega n_1 + \pi i\, {}^t n_1 n_2]\, \vartheta(\Omega n_1 + n_2, \Omega) \ .$$

Written out,

$$(5.3') \qquad \vartheta^\alpha[{}^{n_1}_{n_2}](\Omega) = \sum_{\vec{n}\in\mathbb{Z}^g} \exp[\pi i\, {}^t(n+n_1)\Omega(n+n_1) + 2\pi i\, {}^t\left(n+\frac{n_1}{2}\right)\cdot n_2] \ .$$

If we use $\vartheta^\alpha$ instead of $\vartheta$, the "messy exponential factor" is rather:

$$\begin{aligned} &\pi i({}^t n_1\Omega + {}^t n_2)(C\Omega + D)^{-1} C(\Omega n_1 + n_2) \\ &- \pi i\, {}^t n_1 \Omega n_1 - \pi i\, {}^t n_1 n_2 \\ &+ \pi i({}^t n_1 {}^t D - {}^t n_2 {}^t C)(A\Omega + B)(C\Omega + D)^{-1}(Dn_1 - Cn_2) \\ &+ \pi i({}^t n_1 {}^t D - {}^t n_2 {}^t C).(-Bn_1 + An_2), \end{aligned}$$

which, treated as before, turns out to vanish identically!

Thus the functional equation becomes:

$$(5.4) \qquad \vartheta^\alpha\begin{bmatrix} Dn_1 - Cn_2 \\ -Bn_1 + An_2 \end{bmatrix}((A\Omega + B)(C\Omega + D)^{-1}) = \zeta_\gamma \det(C\Omega + D)^{1/2} \cdot \vartheta^\alpha[{}^{n_1}_{n_2}](\Omega) .$$

How about the factor $\det(C\Omega+D)^{1/2}$? This too can be "eliminated" in a sense. Let

$$\vec{w} = {}^t(C\Omega+D)^{-1}\cdot\vec{z}.$$

Then

$$dw_1\wedge\cdots\wedge dw_g = \det(C\Omega+D)^{-1}\cdot dz_1\wedge\cdots\wedge dz_g.$$

Hence (5.4) says:

$$\vartheta^{\alpha}\begin{bmatrix} Dn_1-Cn_2 \\ -Bn_1+An_2 \end{bmatrix}((A\Omega+B)(C\Omega+D)^{-1}).\ \sqrt{dw_1\wedge\cdots\wedge dw_g}$$

$$= \zeta.\ \vartheta^{\alpha}\begin{bmatrix} n_1 \\ n_2 \end{bmatrix}(\Omega).\ \sqrt{dz_1\wedge\cdots\wedge dz_g}$$

or

<u>Proposition</u> 5.5: <u>Let</u> $Sp(2g,\mathbb{Z})$ <u>act as follows:</u>

a) <u>on</u> $\mathbb{Z}^{2g}$, <u>by</u> $(n_1,n_2) \longmapsto (Dn_1-Cn_2,-Bn_1+An_2)$
b) <u>on</u> $\mathfrak{H}_g$, <u>by</u> $\Omega \longmapsto (A\Omega+B)(C\Omega+D)^{-1}$
c) <u>on</u> $\mathbb{C}^g$, <u>by</u> $\vec{z} \longmapsto {}^t(C\Omega+D)^{-1}\cdot\vec{z}$.

<u>Then the functional equation for</u> $\vartheta$ <u>asserts that, up to an</u> $8^{th}$ <u>root of</u> 1,

$$\vartheta^{\alpha}\begin{bmatrix} n_1 \\ n_2 \end{bmatrix}(\Omega).\ \sqrt{dz_1\wedge\cdots\wedge dz_g}$$

<u>is invariant under</u> $\Gamma_{1,2} \subset Sp(2g,\mathbb{Z})$.

Next, to prove the functional equation, we must consider the 3 generators of $\Gamma_{1,2}$.

<u>Case I</u>: $\gamma = \begin{pmatrix} A & 0 \\ 0 & {}^tA^{-1} \end{pmatrix}$, $A \in GL(g,\mathbb{Z})$.

Then (5.1) reduces to:

$$\vartheta(A\vec{z},\ A.\Omega.{}^tA) = \zeta\ \det(A)^{-1/2}\,\vartheta(\vec{z},\Omega)$$

which is immediate, with $\zeta$ a $4^{th}$ root of 1, in fact:

$$\vartheta(A\vec{z},\ A\Omega{}^tA) = \sum_{n\in\mathbb{Z}^g} \exp[\pi i{}^tnA\Omega{}^tA\cdot n + 2\pi i{}^tn\cdot A\cdot z]$$

$$= \sum_{n\in\mathbb{Z}^g} \exp[\pi i{}^t({}^tAn)\Omega({}^tAn) + 2\pi i{}^t({}^tAn)z]$$

$$= \sum_{m\in\mathbb{Z}^g} \exp[\pi i{}^tm\Omega m + 2\pi i{}^tmz]$$

$$= \vartheta(\vec{z},\Omega)$$

and since $\det A = \pm 1$, $\det(A)^{-1/2}$ is a $4^{th}$ root of 1.

<u>Case II</u>: $\gamma = \begin{pmatrix} I & B \\ 0 & I \end{pmatrix}$, B symmetric, even diagonal.

Then (5.1) reduces to

$$\vartheta(\vec{z},\Omega+B) = \zeta\cdot\vartheta(\vec{z},\Omega).$$

Here we may take $\zeta = +1$, because in fact

$$\vartheta(\vec{z},\Omega+B) = \sum_{n\in\mathbb{Z}^g} \exp[\pi i{}^tn(\Omega+B)n + 2\pi i{}^tnz]$$

$$= \sum \exp[\pi i{}^tnBn]\cdot\exp[\pi i{}^tn\Omega n + 2\pi i{}^tnz]$$

$$= \vartheta(\vec{z},\Omega)$$

because ${}^tnBn$ is always an even integer.

Case III: $\gamma = \begin{pmatrix} 0 & -I \\ I & 0 \end{pmatrix}$.

Then (5.1) reduces to:

(5.6) $$\vartheta(\Omega^{-1}\vec{z}, -\Omega^{-1}) = \zeta\cdot\det(\Omega)^{1/2}\cdot\exp[\pi i\,{}^t z\cdot\Omega^{-1}z]\cdot\vartheta(\vec{z},\Omega).$$

In fact, this is true with $\zeta\cdot\det(\Omega)^{1/2}$ replaced by $\det(\frac{\Omega}{i})^{1/2}$ where the branch of the square root is used which has positive value when $\Omega$ is pure imaginary.

We could prove (5.6) along the lines of the proof of Chapter I, but instead we will give a different proof based on the Poisson Summation Formula:

(5.7) f a smooth function on $\mathbb{R}^g$, going to zero fast enough at $\infty$, $\hat{f}$ its Fourier transform:

$$\hat{f}(\xi) = \int_{\mathbb{R}^g} f(x)\,\exp(2\pi i\,{}^t x.\xi)dx_1\cdots dx_g$$

then

$$\sum_{n\in\mathbb{Z}^g} f(n) = \sum_{n\in\mathbb{Z}^g}\hat{f}(n).$$

We apply this with $f(x) = \exp(\pi i\,{}^t x\Omega x + 2\pi i\,{}^t x\cdot z)$. Then

$$\sum_{n\in\mathbb{Z}^g} f(n) = \vartheta(z,\Omega).$$

To calculate $\hat{f}$, we need the following integral:

Lemma 5.8: For all $\Omega\in\mathfrak{H}_g$, $z\in\mathbb{C}^g$,

$$\int_{\mathbb{R}^g}\exp(\pi i\,{}^t x\Omega x + 2\pi i\,{}^t x.z)dx_1\cdots dx_g = (\det\Omega/i)^{-1/2}\exp(-\pi i\,{}^t z\Omega^{-1}z).$$

Proof: Rewrite the integral as

$$\exp(-\pi i\, {}^t z\Omega^{-1}z) \int_{\mathbb{R}^g} \exp(\pi i\, {}^t(x+\Omega^{-1}z)\Omega(x+\Omega^{-1}z))dx_1,\cdots,dx_g \ .$$

As both sides of the equality to be proved are holomorphic in $\Omega$ and z, it suffices to prove they are equal when $\Omega$ and z are pure imaginary. Therefore, we may assume

$$\Omega = i\,{}^tA\cdot A \quad , \quad A \text{ real positive definite symmetric}$$

$$z = iy \quad , \quad y \text{ real.}$$

Then the integral becomes:

$$\exp(-\pi i\, {}^t z\Omega^{-1}z) \int_{\mathbb{R}^g} \exp\left[-\pi\, {}^t\left(x+({}^tAA)^{-1}y\right){}^tA\cdot A\,\left(x+({}^tAA)^{-1}y\right)\right]dx_1\cdots dx_g.$$

Replacing x by $x+({}^tAA)^{-1}y$, this is

$$\exp(-\pi i\, {}^t z\ \Omega^{-1}z) \int_{\mathbb{R}^g} \exp[-\pi\, {}^tx\,{}^tA\cdot A\ x]dx_1\cdots dx_g.$$

Substituting w = Ax, this becomes

$$\exp(-\pi i\, {}^t z\Omega^{-1}z) \int_{\mathbb{R}^g} \exp[-\pi\, {}^tw\cdot w]\cdot(\det A)^{-1}dw_1\cdots dw_g$$

$$= \exp(-\pi i\, {}^t z\Omega^{-1}z)(\det {}^tA\cdot A)^{-1/2}\cdot \prod_{i=1}^{g} \int_{-\infty}^{+\infty} e^{-\pi w_i^2}\, dw_i$$

$$= \exp(-\pi i\, {}^t z\Omega^{-1}z)(\det \Omega/i)^{-1/2}\ .$$

QED

We may now calculate $\hat{f}$:

$$\hat{f}(\xi) = \int_{\mathbb{R}^g} \exp(\pi i\,{}^t x\Omega x + 2\pi i\,{}^t x\cdot z)\exp(2\pi i\,{}^t x\cdot\xi)\,dx_1\cdots dx_g$$

$$= (\det \Omega/i)^{-1/2}\exp(-\pi i\,{}^t(z+\xi)\Omega^{-1}(z+\xi)).$$

Therefore

$$\sum_{n\in\mathbb{Z}^g} \hat{f}(n) = (\det \Omega/i)^{-1/2}\exp(-\pi i\,{}^t z\Omega^{-1}z)\sum_{n\in\mathbb{Z}^g}\exp(-\pi i\,{}^t n\Omega^{-1}n - 2\pi i\,{}^t n\Omega^{-1}z)$$

$$= (\det \Omega/i)^{-1/2}\exp(-\pi i\,{}^t z\Omega^{-1}z)\,\vartheta(\Omega^{-1}z,-\Omega^{-1})$$

(replacing n by -n in the sum in the last step). This is (5.6).

This completes the proof of the functional equation. A Corollary of our proof which is useful is that:

(5.9) <u>If</u> $\gamma \in \Gamma_4$, i.e., $\gamma \equiv I_{2g}$ (mod 4), <u>then in the functional equation</u>, $\zeta = \pm 1$.

<u>Proof</u>: In fact, by the Appendix, $\Gamma_4$ is contained in the group generated by matrices

$$\begin{pmatrix} I & 2B \\ 0 & I \end{pmatrix}, \begin{pmatrix} I & 0 \\ 2C & I \end{pmatrix}, \quad B,C \text{ symmetric.}$$

For the first of these, the functional equation holds with $\zeta = +1$. But

$$\begin{pmatrix} I & 0 \\ -2B & I \end{pmatrix} = \begin{pmatrix} 0 & -I \\ I & 0 \end{pmatrix}^{-1} \cdot \begin{pmatrix} I & 2B \\ 0 & I \end{pmatrix}\begin{pmatrix} 0 & -I \\ I & 0 \end{pmatrix}$$

so the $8^{th}$ root of unity $\zeta$ involved in the functional equation for $\begin{pmatrix} 0 & -I \\ I & 0 \end{pmatrix}$ cancels out, and $\zeta = \pm 1$ in the equation for $\begin{pmatrix} I & 0 \\ -2B & I \end{pmatrix}$ (we cannot say $\zeta = +1$ unless the appropriate branch of $\sqrt{\det(C\Omega+D)}$ is chosen). <u>QED</u>

We now introduce the general concept of a modular form on $\mathfrak{h}_g$:

Definition 5.10: Let $\Gamma \subset Sp(2g,\mathbb{Z})$ be a subgroup of finite index. Then a modular form of weight k and level $\Gamma$ in g variables is a holomorphic function f on Siegel's upper half-space $\mathfrak{h}_g$ such that for all,

$$\gamma = \begin{pmatrix} A & B \\ C & D \end{pmatrix} \in \Gamma$$

we have

$$f((A\Omega+B)(C\Omega+D)^{-1}) = \det(C\Omega+D)^k \cdot f(\Omega).$$

If $g = 1$, we put an extra boundedness hypothesis on the behaviour of f at the "cusps". If $g > 1$, it turns out that this boundedness is automatic (the "Koecher principle"): for example, f will be bounded in the open set of $\Omega$'s such that

$$\operatorname{Im} \Omega > c.I_g$$

for some constant $c > 0$. If $\Gamma = \Gamma_n$, then f is said to be a modular form of level n. If $g \geq 2$, then by a result of Mennicke (Math. Annalen, 159 (1965), p. 115), any such $\Gamma$ contains some subgroup $\Gamma_n$, so a modular form of level $\Gamma$ is a modular form of level n for some n. The functional equation for $\vartheta$ states then that $\vartheta(0,\Omega)^2$ is a modular form of weight 1 and level 4. More precisely, if we introduce following Igusa the intermediate levels (n,2n) by

$$\Gamma_{2n} \subset \Gamma_{n,2n} \subset \Gamma_n$$

where n is assumed even and $\gamma \in \Gamma_{n,2n}$ if

$$\gamma = \begin{pmatrix} A & B \\ C & D \end{pmatrix} \equiv I_{2g} \bmod n$$

and 2n divides the diagonals of B and C, then we prove:

Corollary 5.11: Let n be even. Then for all $n_1, n_2, m_1, m_2 \in \frac{1}{n}\mathbb{Z}^g$,

$$\vartheta\left[\begin{smallmatrix} n_1 \\ n_2 \end{smallmatrix}\right](0,\Omega)\cdot\vartheta\left[\begin{smallmatrix} m_1 \\ m_2 \end{smallmatrix}\right](0,\Omega)$$

is a modular form of weight 1, level $(n^2, 2n^2)$.

Proof: This follows immediately from (5.2).

In fact, Igusa has shown that the ring of all modular forms of level $(n^2, 2n^2)$, of all integral weights, is just the integral closure of the subring generated by these thetas and that they have the same fraction field. This is the final result in his book Theta functions, Springer, 1972. It is an open problem, however, of considerable interest to understand exactly what subring of the ring of modular forms is generated by the thetas. Geometrically, we can proceed as in Chapter I and define a holomorphic mapping:

$$\left(\mathfrak{h}_g/\Gamma_{(n^2,2n^2)}\right) \longrightarrow \mathbb{P}^{N-1}$$

by

$$\Omega \longmapsto \left(\cdots, \vartheta\left[\begin{smallmatrix} n_1^\alpha \\ n_2^\alpha \end{smallmatrix}\right](0,\Omega)\cdot\vartheta\left[\begin{smallmatrix} m_1^\alpha \\ m_2^\alpha \end{smallmatrix}\right](0,\Omega), \cdots\right)$$

where $\left(\begin{smallmatrix} n_1^\alpha \\ n_2^\alpha \end{smallmatrix}\right), \left(\begin{smallmatrix} m_1^\alpha \\ m_2^\alpha \end{smallmatrix}\right)$ runs through all sets of 4 elements in a system of coset representatives of $\frac{1}{n}\mathbb{Z}^g$ modulo $\mathbb{Z}^g$. The main result geometrically is that this is an isomorphism of the analytic space $\mathfrak{h}_g/\Gamma_{(n^2,2n^2)}$ with a "quasi-projective" variety, i.e., a subset of $\mathbb{P}^{N-1}$ defined by polynomial equations minus a smaller set of the same type.

Corollary 5.11 suggests that we extend the definition of modular forms to half-integral weights as follows:

Definition 5.12. Let $\Gamma \subset \Gamma_{1,2}$ be a subgroup of finite index. Then a modular form f of weight $k \in \frac{1}{2}\mathbb{Z}$ and level $\Gamma$ is a holomorphic function f on $\mathfrak{H}_g$ such that

$$\frac{f((A\Omega+B)(C\Omega+D)^{-1})}{\vartheta[{0 \atop 0}]((A\Omega+B)(C\Omega+D)^{-1})^{2k}} = \frac{f(\Omega)}{\vartheta[{0 \atop 0}](\Omega)}$$

for all $\begin{pmatrix} A & B \\ C & D \end{pmatrix} \in \Gamma$.

With this definition, we can even extend 5.11 as follows:

Corollary 5.13: For all $n_1, n_2 \in \mathbb{Q}^g$, $\ell \in \mathbb{Z}, \ell \geq 1$,

$$\vartheta\begin{bmatrix} n_1 \\ n_2 \end{bmatrix}(0, \ell\Omega)$$

is a modular form of weight 1/2 for a suitable level $\Gamma$.

Proof: Consider

$$f(\Omega) = \vartheta\begin{bmatrix} n_1 \\ n_2 \end{bmatrix}(0,\ell\Omega) \Big/ \vartheta[{0 \atop 0}](0,\Omega).$$

Substituting $(A\Omega+B)(C\Omega+D)^{-1}$ for $\Omega$ and using (5.2), it follows that if $n_1, n_2 \in 1/n\mathbb{Z}^g$, n even, and

$$\begin{pmatrix} A & \ell B \\ \ell^{-1}C & D \end{pmatrix} \in \Gamma_{n^2,2n^2}$$

then

$$f((A\Omega+B)(C\Omega+D)^{-1}) = \pm f(\Omega).$$

The sign $\varepsilon\begin{pmatrix} A & B \\ C & D \end{pmatrix}$ gives a homomorphism

$$\Gamma_{1,2} \cap \left[\begin{pmatrix} \ell & 0 \\ 0 & 1 \end{pmatrix}^{-1} \cdot \Gamma_{n^2,2n^2} \cdot \begin{pmatrix} \ell & 0 \\ 0 & 1 \end{pmatrix}\right] \longrightarrow \{\pm 1\} .$$

Let $\Gamma$ be the kernel. Then $\Gamma$ is a level for $\vartheta\begin{bmatrix} n_1 \\ n_2 \end{bmatrix}(0,\ell\Omega)$. QED

Another way to describe the situation is this:

let

$$\mathcal{S}(\mathbb{Q}^g) = \left\{\begin{array}{l} \text{vector space of functions } f: \mathbb{Q}^g \longrightarrow \mathbb{C} \\ \text{such that for some } k,\ell \geq 1 \\ \qquad f(\vec{a}) = 0 \text{ if } \vec{a} \notin \frac{1}{k}\mathbb{Z}^g \\ \qquad f(\vec{a}+\vec{b}) \equiv f(\vec{a}) \text{ if } \vec{b} \in \ell\,\mathbb{Z}^g \quad . \end{array}\right\}$$

($\mathcal{S}$ is also called the space of Schwartz functions on the group $\mathbb{A}_f^g$, $\mathbb{A}_f$ being the finite adèles). Define

$$\vartheta[f](\ell\Omega) = \sum_{\vec{n}\in\mathbb{Q}^g} f(\vec{n})\exp(\pi i\,{}^t\vec{n}\Omega\vec{n}) \quad .$$

Then

$$f \longmapsto \vartheta[f](\ell\Omega)$$

is a map

$$w_\ell: \mathcal{S}(\mathbb{Q}^g) \longrightarrow \{\text{v.sp. of modular forms of wt. } \tfrac{1}{2}, \text{ any level}\}.$$

The image is the same as the span of the modular forms $\vartheta[\begin{smallmatrix} n_1 \\ n_2 \end{smallmatrix}](0,\ell\Omega)$, all $n_1,n_2 \in \mathbb{Q}^g$ because $\vartheta[f]$ becomes $\vartheta[\begin{smallmatrix} n_1 \\ n_2 \end{smallmatrix}]$ if f is taken to be the characteristic function of $\vec{a}+\mathbb{Z}^g$ times the character defined by $\vec{b}$. $w_\ell$ is known as "the Weil map associated to the 1-variable quadratic form $\ell x^2$".

## Appendix to §5: Generators of $Sp(2g,\mathbb{Z})$

In the last section and in the next Chapter, we need at various places lemmas asserting that various subgroups of $Sp(2g,\mathbb{Z})$ of finite index are generated by such and such elements. We group together all the results of this ilk that we need. There is nothing very difficult in any of these. First, the subgroups we shall consider are:

$$\Gamma_n = \{\gamma \in Sp(2g,\mathbb{Z}) \mid \gamma \equiv I_{2g} \bmod n\}$$

and also an intermediate subgroup $\Gamma_{1,2}$:

$$\Gamma_2 \subset \Gamma_{1,2} \subset \Gamma_1 = Sp(2g,\mathbb{Z}) .$$

In fact, if $Sp(2g,\mathbb{Z})$ acts by reduction mod 2 on $(\mathbb{Z}/2\mathbb{Z})^{2g}$, it preserves the skew-symmetric form

$$A((x_1,x_2),(y_1,y_2)) = {}^t x_1 \cdot y_2 - {}^t x_2 \cdot y_1$$

which, because the characteristic is 2, is also symmetric. In fact, over $\mathbb{Z}/2\mathbb{Z}$, consider the quadratic form

$$Q((x_1,x_2)) = {}^t x_1 \cdot x_2 .$$

Then

$$A(x,y) \equiv Q(x+y) - Q(x) - Q(y) \pmod 2 .$$

Therefore, the orthogonal group over $\mathbb{Z}/2\mathbb{Z}$ (the maps preserving Q) is a subgroup of the symplectic group over $\mathbb{Z}/2\mathbb{Z}$ (the maps preserving A)! Let

$$\Gamma_{1,2} = \{\gamma \in Sp(2g,\mathbb{Z}) \mid Q(\gamma x) \equiv Q(x) \bmod 2\} .$$

We rest our sequence of generation assertions on one dealing with the fewest generators:

Proposition A.1: Let $\Gamma \subset Sp(2g,\mathbb{Z})$ be the subgroup generated by the elements

$$\begin{pmatrix} I & 2B \\ 0 & I \end{pmatrix}, \begin{pmatrix} I & 0 \\ 2C & I \end{pmatrix}, \quad B,C \text{ integral, symmetric}.$$

Then $\Gamma_4 \subset \Gamma \subset \Gamma_2$. In fact $\Gamma$ is the group $\tilde{\Gamma}$ of $\begin{pmatrix} A & B \\ C & D \end{pmatrix}$ where $4|A-I_g$, $4|D-I_g$, $2|B$, $2|C$.

Proof: We use

Lemma A.2: Let $n,m \in \mathbb{Z}$, not both zero. Let $d = (n,m)$. Then a sequence of the elementary transformations

$$(x,y) \longmapsto (x\pm 2y,y) \quad \text{and} \quad (x,y\pm 2x)$$

carries $(n,m)$ to either

$$(d,0),(-d,0),(0,d),(0,-d) \text{ or } (d,d).$$

Proof: Since everything preserves divisibility by d, we may as well divide by d and prove this for $d = 1$. Given $(n,m)$, either $|n| < |m|$, $|n| > |m|$ or $|n| = |m|$. If $0 \neq |n| < |m|$, make the map

$$(x,y) \longmapsto (x,y+2x) \text{ or } (x,y-2x)$$

so as to decrease $|m|$. If $|n| > |m| \neq 0$, make the map

$$(x,y) \longmapsto (x+2y,y) \text{ or } (x-2y,y)$$

so as to decrease $|n|$. If $|n| = |m|$, then $n = \pm m$ and since $(n,m) = 1$, $|n| = |m| = 1$. If $(n,m) = (-1,1)$ or $(1,-1)$, one of the elementary transformations carries it to $(1,1)$. If $(n,m) = (-1,-1)$, we need 2 of them:

$$(-1,-1) \longmapsto (-1,+1) \longmapsto (+1,+1)$$

$$(x,y) \longmapsto (x,y-2x)$$

$$(x,y) \longmapsto (x+2y,y). \qquad \underline{\text{QED}}$$

To prove the Proposition, let $\tilde{\gamma} \in \tilde{\Gamma}$. Consider

$$\tilde{\gamma}(1,0,\cdots,0;0,\cdots,0)) = (a_1,\cdots,a_g;b_1,\cdots,b_g).$$

Here $4|a_1-1,a_2,\cdots,a_g, 2\Big|b_1,\cdots,b_g$ and g.c.d.$(a_1,\cdots,b_g) = 1$. We shall follow $\tilde{\gamma}$ by a sequence $\delta_1,\cdots,\delta_N$ of elementary transformations $\begin{pmatrix} I & 2B \\ 0 & I \end{pmatrix}, \begin{pmatrix} I & 0 \\ 2C & I \end{pmatrix}$ until

$$\delta_N\delta_{N-1}\cdots\ \delta_1\tilde{\gamma}(1,0,\cdots,0;0,\cdots,0) = (1,0,\cdots,0;0,\cdots,0).$$

Note that we may have at our disposal the transformations:

$$a_i,b_i \longmapsto a_i\pm 2b_i,b_i \qquad \text{, other } a_k,b_k \text{ left alone}$$
$$a_i,b_i \longmapsto a_i,b_i\pm 2a_i \qquad \text{,} \quad " \quad "$$
$$a_i,a_j,b_i,b_j \longmapsto a_i\pm 2b_j,a_j\pm 2b_i,b_i,b_j, \quad " \quad "$$
$$a_i,a_j,b_i,b_j \longmapsto a_i,a_j,b_i\pm 2a_j,b_j\pm 2a_i, \quad " \quad "$$

We proceed in stages like this:

<u>Step I</u>: Let $d = (a_1,b_1)$. Note that d is odd because $a_1$ is odd. Apply $(a_1,b_1) \longmapsto (a_1\pm 2b_1,b_1)$ or $(a_1,b_1\pm 2a_1)$. By the lemma, we eventually achieve

$$\underline{a_1 = \pm d}, \quad \underline{b_1 = 0} \quad .$$

(The other possibilities are excluded because d is odd and $b_1$ always remains even at each stage.)

Step II: We want to decrease $|a_1|$. If $a_1 \nmid b_i$ for some i, we apply

$$a_1, a_i, b_1, b_i \longmapsto a_1 \pm 2b_i, a_i \pm 2b_1, b_1, b_i$$

$$\text{or} \quad a_1, a_i, b_1 \pm 2a_i, b_i \pm 2a_1 .$$

Again because $a_1$ is odd, $b_i$ is even, if $d' = \text{g.c.d.}(a_1, b_i)$, we eventually reach

$$a_1 = \pm d', \ b_i = 0.$$

Step III: Repeat Step II until $a_1 | b_i$, all i. We also may repeat $a_1, b_1 \longmapsto a_1, b_1 \pm 2a_1$ until $b_1 = 0$ again. We want to decrease $|a_1|$ further. If $a_1 \nmid a_i$ $(i \geq 2)$, we first apply

$$a_1, a_i, b_1, b_i \longmapsto a_1, a_i, b_1 + 2a_i, b_i + 2a_1$$

so that $b_1$ becomes $2a_i$, then repeat Step I. In this way, we decrease $|a_1|$ until $a_1 | a_i, b_i$. Then as $\text{g.c.d.}(a_i, b_i) = 1$, $a_1 = \pm 1$.

Step IV: Kill $b_2, \cdots, b_g$ by maps

$$a_1, a_i, b_1, b_i \longmapsto a_1, a_i, b_1 \pm 2a_i, b_i \pm 2a_1.$$

Make $b_1 = 2$ by maps

$$a_1, b_1 \longmapsto a_1, b_1 \pm 2a_1.$$

Step V: Kill $a_2, \cdots, a_g$ by maps

$$a_1, a_i, b_1, b_i \longmapsto a_1 \pm 2b_i, a_i \pm 2b_1, b_1, b_i.$$

(These don't affect $a_1$ because $b_i = 0$, $i > 1$; and since $4 | a_i$, $i \geq 2$, $a_i$ is a multiple of $2b_1$.) Finally kill $b_1$ by

$$a_1, b_1 \longmapsto a_1, b_1 - 2a_1.$$

Next, consider

$$\delta_N \delta_{N-1} \cdots \delta_1 \tilde{\gamma}(0, \cdots, 0; 1, 0, \cdots, 0) = (c_1, \cdots, c_g, d_1, \cdots, d_g).$$

Because $\delta_N \cdots \tilde{\gamma}$ is symplectic, and maps $(1, \cdots, 0)$ to $(1, \cdots, 0)$, we must have $d_1 = 1$. Moreover, as $\delta_N \cdots \tilde{\gamma} \in \tilde{\Gamma}$, we have $2 | c_1, \cdots, c_g$, $4 | d_2, \cdots, d_g$. We choose more elementary transformations $\delta_i$ until

$$\delta_M \cdot \delta_{M-1} \cdots \delta_N \delta_{N-1} \cdots \delta_1 \tilde{\gamma}$$

fixes $(1, \cdots, 0)$ <u>and</u> $(0, \cdots, 0; 1, \cdots, 0)$.

<u>Step VI</u>: Kill $c_2, \cdots, c_g$ and make $c_1 = 2$ by maps

$$(c_1, c_i, d_1, d_i) \longmapsto (c_1 \pm 2d_i, c_i \pm 2d_1, d_1, d_i).$$

<u>Step VII</u>: Kill $d_2, \cdots, d_g$ by maps

$$(c_1, c_i, d_1, d_i) \longmapsto (c_1, c_i, d_1 \pm 2c_i, d_i \pm 2c_1)$$

and finally kill $c_1$ by

$$(c_1, d_1) \longmapsto (c_1 - 2d_1, d_1).$$

The Proposition now follows by induction on $g$, because $(\delta_M \cdots \tilde{\gamma})$ preserves the direct sum decomposition

$$\mathbb{Z}^{2g} = \begin{bmatrix}\mathbb{Z}(1,0,\cdots,0;0,\cdots,0) \\ +\ \mathbb{Z}(0,\cdots,0;1,\cdots,0)\end{bmatrix} \oplus \left[(n_1,n_2) \,\middle|\, (n_1)_1 = (n_2)_1 = 0\right]$$

and is the identity on the first piece. On the $2^{nd}$ piece, we have an element of $Sp(2g-2,\mathbb{Z})$. QED

There are 2 useful ways to get generators of $\Gamma_2$:

Proposition A3: $\Gamma_2$ is generated by either of the following:

a) $\begin{pmatrix} I & 2B \\ 0 & I \end{pmatrix}, \begin{pmatrix} I & 0 \\ 2C & I \end{pmatrix}, \begin{pmatrix} A & 0 \\ 0 & {}^tA^{-1} \end{pmatrix}$, where $A \equiv I_g \pmod 2$

or

b) the transformations

b$_1$) $\vec{x} \longmapsto \vec{x}+2A(\vec{x},\vec{e}_i)\cdot\vec{e}_i$, $\quad 1 \le i \le 2g$

b$_2$) $\vec{x} \longmapsto \vec{x}+2A(\vec{x},\vec{e}_i+\vec{e}_j)(\vec{e}_i+\vec{e}_j)$, $\quad 1 \le i < j \le 2g$

where $\vec{e}_i \in \mathbb{Z}^{2g}$ are the unit vectors.

Proof: Both of these contain the generators of A1 (in (b), use the maps $b_1$ and $b_2$ with $1 \le i < j \le g$ and $g+1 \le i < j \le 2g$), hence generate a subgroup containing $\tilde{\Gamma}$. On the other hand, $\Gamma_2/\Gamma_4$ is an abelian group, which may be described as

$$\Gamma_2/\Gamma_4 = I_{2g} + 2\left\{\begin{pmatrix} A_0 & B_0 \\ C_0 & D_0 \end{pmatrix} \bmod 2 \,\middle|\, B_0, C_0 \text{ symmetric, } D_0 = -{}^tA_0\right\}.$$

(Check this by examining the condition

$$A((I+2X)x,(I+2X)y) = A(x,y)$$

modulo 4). It suffices to check that the generators in (a) and (b) generate $\Gamma_2/\tilde{\Gamma}$, i.e., contain elements

$$\equiv \begin{pmatrix} I+2A_0 & * \\ * & I-2\,{}^tA_0 \end{pmatrix} \bmod 4$$

for every $g\times g$ integral $A_0$. In (a), take upper and lower triangular A's with $\pm 1$'s on the diagonal. In (b), we put the maps $b_2$ in matrix form. Thus if $i = 1$, $j = g+1$, it is

$$\left(\begin{array}{ccc|ccc} 3 & & 0 & -2 & & 0 \\ & 1 \ddots & & & & \\ 0 & & 1 & 0 & & 0 \\ \hline 2 & & 0 & -1 & & 0 \\ & & & & 1 \ddots & \\ 0 & & 0 & 0 & & 1 \end{array}\right)$$

so that if $j = g+i$, we get diagonal $A_0$'s. And if $i = 1$, $j = g+2$, we get

$$\left(\begin{array}{cc|c|cc|c} 1 & -2 & \multirow{2}{*}{0} & 2 & 0 & \multirow{2}{*}{0} \\ 0 & 1 & & 0 & 0 & \\ \hline \multicolumn{2}{c|}{0} & 1 \ddots 1 & \multicolumn{2}{c|}{0} & 0 \\ \hline 0 & 0 & \multirow{2}{*}{0} & 1 & 0 & \multirow{2}{*}{0} \\ 0 & -2 & & 2 & 1 & \\ \hline \multicolumn{2}{c|}{0} & 0 & \multicolumn{2}{c|}{0} & 1 \ddots 1 \end{array}\right)$$

These give off-diagonals $A_0$'s. QED

Proposition A 4. $\Gamma_{1,2}$ is generated by

$$\begin{pmatrix} 0 & I \\ -I & 0 \end{pmatrix}, \quad \begin{pmatrix} A & 0 \\ 0 & {}^tA^{-1} \end{pmatrix}, \quad \begin{pmatrix} I & B \\ 0 & I \end{pmatrix}$$

all $A \in GL(g,\mathbb{Z})$, symmetric integral $B$ with even diagonal.

Proof: It suffices by A 3 to prove that $\Gamma_{1,2}/\Gamma_2$ is generated by images of these elements. This means that we need only show that an orthogonal map of $(\mathbb{Z}/2\mathbb{Z})^{2g}$ to $(\mathbb{Z}/2\mathbb{Z})^{2g}$ is composed of maps:

a) $x_1,\cdots,x_g,y_1,\cdots,y_g \longmapsto y_1,\cdots,y_g,x_1,\cdots,x_g$

b) $x_i,x_j,y_i,y_j \longmapsto x_i,x_j+x_i,y_i+y_j,y_j$ other $x_k,y_k$ fixed

c) $x_i,x_j,y_i,y_j \longmapsto x_i,x_j,y_i+x_j,y_j+x_i$ "

d) $x_i,x_j,y_i,y_j \longmapsto x_i+y_j,x_j+y_i,y_i,y_j$ "

This can be done exactly as in the proof of A1. Let $\tilde{\gamma}$ be an orthogonal map and say

$$\tilde{\gamma}(1,0,\cdots,0;0,\cdots,0) = (a_1,\cdots,a_g,b_1,\cdots,b_g) \quad \text{(note: } a_i,b_i\text{'s are 0 or 1).}$$

First, use map (a) to ensure that not all $a_i$'s are 0. Use maps (b) to make only one $a_i$ equal to 1, and then to make in fact $a_1 = 1$, $a_2 =\cdots= a_g = 0$. Use map (c) to make $b_2 =\cdots = b_g = 0$. Then because

$$Q(1,0,\cdots,0;0,\cdots,0) = 0$$

we have $Q(a_1,\cdots,a_g,b_1,\cdots,b_g) = 0$ too, so in fact at this stage $b_1$ must be zero too, i.e.,

$$\delta_N\delta_{N-1}\cdots\delta_1\tilde{\gamma}(1,0,\cdots,0;0,\cdots,0) = (1,0,\cdots,0;0,\cdots,0).$$

Next look at

$$\delta_N\cdots\tilde{\gamma}(0,\cdots,0;1,0,\cdots,0) = (c_1,\cdots,c_g;d_1,\cdots,d_g).$$

Because its inner product with $(1,\cdots,0;0,\cdots,0)$ is 1, $d_1 = 1$. Use maps (b) to kill $d_2,\cdots,d_g$ and maps (d) to kill $c_2,\cdots,c_g$, while not moving $(1,0,\cdots,0;0,\cdots,0)$. Then because $Q(c_1,\cdots,c_g;d_1,\cdots,d_g) = 0$, we find $c_1 = 0$ too. Thus

$$\delta_M\delta_{M-1}\cdots\delta_N\delta_{N-1}\cdots\tilde{\gamma}$$

fixes $(1,\cdots,0;0,\cdots,0)$ *and* $(0,\cdots,0;1,\cdots,0)$. As in A1, using induction, this proves the result. *QED*

Finally:

*Proposition* A5. $Sp(2g,\mathbb{Z})$ *is generated by*

$$\begin{pmatrix} 0 & I \\ -I & 0 \end{pmatrix}, \begin{pmatrix} A & 0 \\ 0 & {}^tA^{-1} \end{pmatrix}, \begin{pmatrix} I & B \\ 0 & I \end{pmatrix}$$

*all* $A \in GL(g,\mathbb{Z})$, B *symmetric, integral*.

*Proof*: We prove this *exactly* as we proved A4, except that at the 2 points where we used the invariance of Q, we use instead maps

e) $x_i,y_i \longmapsto x_i,y_i+x_i$, other $x_k,y_k$ fixed

or f) $x_i,y_i \longmapsto x_i+y_i,y_i$, "

derived from diagonal B's. *QED*

## §6. Riemann's Theta formula and theta functions associated to a quadratic form.

We have described how the functions $\vartheta\left[\begin{smallmatrix} n_1 \\ n_2 \end{smallmatrix}\right](z,\Omega)$ can be used

i) for fixed $\Omega$, $z$ variable, to embed complex tori in $\mathbb{P}^N$

ii) for $z = 0$, variable, to embed $\mathfrak{H}_g/\Gamma_*$ in $\mathbb{P}^N$.

Since these maps are not surjective, there must be polynomial identities between the various functions $\vartheta\left[\begin{smallmatrix} n_1 \\ n_2 \end{smallmatrix}\right](z,\Omega)$. With only a few exceptions, all identities that I know of are deduced from the theta identities of Riemann. These are generalizations of the Riemann identity given in Ch. I for the one-variable case. We will conclude this chapter by describing these.

We start with any rational orthogonal $h\times h$ matrix $T$.

**Theorem 6.1.** (Generalized Riemann theta identity):

$$(R^T) \quad \prod_{i=1}^{h} \vartheta\Big(\sum_{j=1}^{h} t_{ij}\vec{z}_j\Big) = d^{-g} \sum_{A,B\in K} \exp\left[\pi i \operatorname{tr}({}^t\!A\Omega A + 2\,{}^t\!A(z+B))\right] \prod_{i=1}^{h} \vartheta(\vec{z}_i + \Omega\alpha_i + \beta_i)$$

($z,\alpha,\beta$ $g\times 1$ column vectors, $A = (\alpha_1,\ldots,\alpha_h)$, $B = (\beta_1,\ldots,\beta_h)$ $g\times h$ matrices, $\mathbb{Z}^{(g,h)}$ = group of integral $g\times h$ matrices,

$$K = \mathbb{Z}^{(g,h)}\cdot T/\mathbb{Z}^{(g,h)}\cdot T\cap\mathbb{Z}^{(g,h)}, \quad \text{and} \quad d = [T^{-1}\mathbb{Z}^h : T^{-1}\mathbb{Z}^h\cap\mathbb{Z}^h]).$$

The main example is: $h = 4$, $T = \frac{1}{2}\begin{pmatrix} 1 & 1 & 1 & 1 \\ 1 & 1 & -1 & -1 \\ 1 & -1 & 1 & -1 \\ 1 & -1 & -1 & 1 \end{pmatrix}$, so that

$T = {}^tT = T^{-1} = {}^tT^{-1}$. Note that

$$T\mathbb{Z}^4 = \left\{ (a_1,a_2,a_3,a_4) \;\middle|\; \begin{array}{l} a_i \in \frac{1}{2}\mathbb{Z} \\ a_i + a_j \in \mathbb{Z} \\ a_1+a_2+a_3+a_4 \in 2\mathbb{Z} \end{array} \right\}$$

so that coset representatives for $T\mathbb{Z}^4 \cap \mathbb{Z}^4$ in $T\mathbb{Z}^4$ are $(0,0,0,0)$ and $(\frac{1}{2},\frac{1}{2},\frac{1}{2},\frac{1}{2})$; the identity becomes

$$\text{(R)} \qquad \vartheta\left(\frac{x+y+u+v}{2}\right)\vartheta\left(\frac{x+y-u-v}{2}\right)\vartheta\left(\frac{x-y+u-v}{2}\right)\vartheta\left(\frac{x-y-u+v}{2}\right)$$

$$= 2^{-g} \sum_{\beta \in \frac{\frac{1}{2}\mathbb{Z}^g}{\mathbb{Z}^g}} \sum_{\alpha \in \frac{\frac{1}{2}\mathbb{Z}^g}{\mathbb{Z}^g}} \exp[4\pi i\,{}^t\alpha\Omega\alpha + 2\pi i\,{}^t\alpha(x+y+u+v)]\,\vartheta(x+\Omega\alpha+\beta)\,\vartheta(y+\Omega\alpha+\beta)\,\vartheta(u+\Omega\alpha+\beta)\,\vartheta(v+\Omega\alpha+\beta)$$

The exponential factor simplifies if we use $\vartheta$-functions with characteristics:

$$(R^T_{ch}) \qquad \prod_{i=1}^{h} \vartheta\begin{bmatrix} \sum_{j=1}^{h} t_{ij}\gamma_j \\ \sum_{j=1}^{h} t_{ij}\delta_j \end{bmatrix}\left(\sum_{j=1}^{h} t_{ij}z_j\right) =$$

$$= [T^{-1}\mathbb{Z}^h : T^{-1}\mathbb{Z}^h \cap \mathbb{Z}^h]^{-g} \cdot \sum_{\beta_1,\ldots,\beta_h} \sum_{\alpha_1,\ldots,\alpha_h} \exp\left(-2\pi i \sum_i {}^t\beta_i \cdot \gamma_i\right) \prod_{i=1}^{h} \vartheta\begin{bmatrix} \alpha_i + \gamma_i \\ \beta_i + \delta_i \end{bmatrix}(z_i).$$

We can derive this from (6.1) as follows:

$$\prod_i \vartheta\begin{bmatrix}\Sigma t_{ij}\gamma_j\\ \Sigma t_{ij}\delta_j\end{bmatrix}(\Sigma t_{ij}z_j)$$

$$= \prod_i \exp\Big(\pi i\,{}^t(\Sigma t_{ij}\gamma_j)\Omega(\Sigma t_{ij}\gamma_j)+2\pi i\,{}^t(\Sigma t_{ij}\gamma_j)(\Sigma t_{ij}(z_j+\delta_j))\Big)\cdot$$

$$\cdot\,\vartheta(\Sigma t_{ij}(z_j+\Omega\gamma_j+\delta_j))$$

$$= \exp\Big(\pi i\ \mathrm{tr}[{}^tC\Omega C+2\,{}^tC(Z+D)]\Big)\prod_i \vartheta(\sum_j t_{ij}(z_j+\Omega\gamma_j+\delta_j))$$

because ${}^tT\cdot T = I$; now apply the theorem to $\vartheta(\sum_j t_{ij}\ (z_j+\Omega\gamma_j+\delta_j))$;

$$= [\quad]^{-g}\sum_A\sum_B \exp\Big(\pi i\ \mathrm{tr}[{}^tC\Omega C+2\,{}^tC(Z+D)]\Big)\exp\Big(\pi i\ \mathrm{tr}[{}^tA\Omega A+2\,{}^tA(Z+\Omega C+D+B)]\Big).$$

$$\prod_i \vartheta(z_i+\Omega\gamma_i+\delta_i+\Omega\alpha_i+\beta_i)$$

$$= [\quad]^{-g}\sum_A\sum_B \exp\Big(\pi i\ \mathrm{tr}({}^t(A+C)\ \Omega(A+C)+2\,{}^t(A+C)(Z+B+D)\Big)\exp\Big(-\pi i\,\mathrm{tr}(2\,{}^tC\cdot B)\Big)\prod_i \vartheta(\ldots\ldots)$$

$$= [\quad]^{-g}\sum_A\sum_B \exp\Big(-2\pi i\ \mathrm{tr}\ {}^tC\cdot B\Big)\prod_i \vartheta\begin{bmatrix}\alpha_i+\gamma_i\\ \beta_i+\delta_i\end{bmatrix}(z_i)\ .$$

Thus in the main example

$$\vartheta\begin{bmatrix}\frac{a+b+c+d}{2}\\ \frac{e+f+g+h}{2}\end{bmatrix}\left(\frac{x+y+u+v}{2}\right)\cdots\vartheta\begin{bmatrix}\frac{a-b-c+d}{2}\\ \frac{e-f-g+h}{2}\end{bmatrix}\left(\frac{x-y-u+v}{2}\right) =$$

$(R_{ch})$

$$2^{-g}\sum_{\alpha,\beta\in\frac{1}{2}\mathbb{Z}^g/\mathbb{Z}^g}\exp\left(-2\pi i\,{}^t\beta(a+b+c+d)\right)\cdot\vartheta\begin{bmatrix}a+\alpha\\ e+\beta\end{bmatrix}(x)\cdots\vartheta\begin{bmatrix}d+\alpha\\ h+\beta\end{bmatrix}(v).$$

This is the formula used in most applications.

Proof of the Theorem: Let $Z = (\vec{z}_1,\cdots,\vec{z}_h) \in \mathbb{C}^{(g,h)}$ be a complex $g\times h$ matrix variable. Then

$$\text{LHS} = \sum_{\vec{n}_1,\cdots,\vec{n}_h\in\mathbb{Z}^g}\exp\left(\pi i\sum_i {}^t\vec{n}_i\Omega\vec{n}_i + 2\pi i\sum_{i,j}({}^t\vec{n}_i\cdot\vec{z}_j)\cdot t_{ij}\right)$$

$$= \sum_{N\in\mathbb{Z}^{(g,h)}}\exp\left(\pi i\,\mathrm{tr}({}^tN\Omega N) + 2\pi i\,\mathrm{tr}({}^tN\cdot Z\cdot{}^tT)\right)$$

$$= \sum_{N\in\mathbb{Z}^{(g,h)}}\exp\left(\pi i\,\mathrm{tr}({}^t(NT)\cdot\Omega\cdot NT) + 2\pi i\,\mathrm{tr}({}^t(N\cdot T)\cdot Z)\right)$$

$$= \sum_{M\in\mathbb{Z}^{(g,h)}\cdot T}\exp\left(\pi i\,\mathrm{tr}({}^tM\Omega M) + 2\pi i\,\mathrm{tr}({}^tM\cdot Z)\right).$$

$$\text{RHS} = [\ \ ]^{-g} \sum_{\vec\beta_1,\dots,\vec\beta_h} \sum_{\vec\alpha_1,\dots,\vec\alpha_h} \sum_{\vec n_1,\dots,\vec n_h} \exp\left(\pi i \sum_i {}^t\vec\alpha_i \Omega \vec\alpha_i + 2\pi i \sum_i {}^t\vec\alpha_i (\vec z_i + \vec\beta_i)\right).$$

$$\exp\left(\pi i \sum_i {}^t\vec n_i \Omega \vec n_i + 2\pi i \sum_i {}^t\vec n_i (\vec z_i + \Omega\vec\alpha_i + \vec\beta_i)\right)$$

$$= [\ \ ]^{-g} \sum_{\vec\beta_i} \sum_{\vec\alpha_i} \sum_{\vec n_i} \exp\left(\pi i \sum_i {}^t(\vec n_i + \vec\alpha_i)\Omega(\vec n_i + \vec\alpha_i) + 2\pi i \sum_i {}^t(\vec n_i + \vec\alpha_i)(\vec z_i + \vec\beta_i)\right)$$

$$= [\ \ ]^{-g} \sum_B \sum_A \sum_{N \in \mathbb{Z}^{(g,h)}} \exp\left(\pi i\, \mathrm{tr}({}^t(N+A)\Omega(N+A)) + 2\pi i\, \mathrm{tr}({}^t(N+A)\cdot(Z+B))\right)$$

where A and B are summed over $\mathbb{Z}^{(g,h)}\cdot T / \mathbb{Z}^{(g,h)}\cdot T \cap \mathbb{Z}^{(g,h)}$. Collecting the sum over A and N together, we get

$$= [\ \ ]^{-g} \sum_{B \in \frac{\mathbb{Z}^{(g,h)}\cdot T}{\mathbb{Z}^{(g,h)}\cdot T \cap \mathbb{Z}^{(g,h)}}} \sum_{M \in (\mathbb{Z}^{(g,h)}\cdot T + \mathbb{Z}^{(g,h)})} \exp\left(\pi i\, \mathrm{tr}({}^tM\Omega M) + 2\pi i\, \mathrm{tr}({}^tM\cdot Z)\right)\cdot$$

$$\cdot \exp[2\pi i\, \mathrm{tr}\, {}^tM\cdot B]\ .$$

Note that

$$B \longmapsto \exp[2\pi i\, \mathrm{tr}\, {}^tM\cdot B]$$

is a character of $M \in \mathbb{Z}^{(g,h)}\cdot T + \mathbb{Z}^{(g,h)}$ trivial on $\mathbb{Z}^{(g,h)}\cdot T$, and all such characters occur for some B. Thus

$$\sum_B \exp(2\pi i\, \mathrm{tr}\, {}^tM\cdot B) = 0$$

unless $\mathrm{tr}({}^tM\cdot B) \in \mathbb{Z}$ for all $B \in \mathbb{Z}^{(g,h)}\cdot T$, i.e., unless $M \in \mathbb{Z}^{(g,h)}\cdot T$. If $M \in \mathbb{Z}^{(g,h)}\cdot T$, all these characters are 1 and we get the order of $\mathbb{Z}^{(g,h)}\cdot T/\mathbb{Z}^{(g,h)}\cdot T \cap \mathbb{Z}^{(g,h)}$. Thus the sum reduces to:

$$= \sum_{M\in\mathbb{Z}^{(g,h)}\cdot T} \exp\Big(\pi i\ \mathrm{tr}({}^tM\Omega M) + 2\pi i\ \mathrm{tr}({}^tM\cdot Z)\Big)$$

$$= \text{LHS}.$$

QED

The reader will notice that the proof of Riemann's theta relation is much shorter than the statement and its rearrangement into its various forms! In fact, as often happens in such a case, if we generalize it even further, the proof will become really simple and transparent. The natural setting to which these ideas lead us is that of theta functions associated to quadratic forms. It is in this setting that all the multiplicative properties of theta functions are best studied. To start, suppose we decide to rewrite a product of h theta functions as one series. What happens is this:

$$\prod_{i=1}^{h} \vartheta(\vec{z}_i,\Omega) = \sum_{\vec{n}_1,\ldots,\vec{n}_h\in\mathbb{Z}^g} \exp\Big(\sum_{i=1}^{h}(\pi i\,{}^t\vec{n}_i\Omega\vec{n}_i + 2\pi i\,{}^t\vec{n}_i\cdot\vec{z}_i)\Big)\ .$$

In terms of

$N = (\vec{n}_1,\ldots,\vec{n}_h)$ be the $g\times h$ matrix with columns $\vec{n}_i$

$Z = (\vec{z}_1,\ldots,\vec{z}_2)$ " " " " " " $\vec{z}_i$ .

This is

$$\prod_{i=1}^{h} \vartheta(\vec{z}_i, \Omega) = \sum_{n\in\mathbb{Z}^{(g,h)}} \exp(\pi i\ \mathrm{tr}({}^t N\cdot\Omega\cdot N) + 2\pi i\ \mathrm{tr}({}^t N\cdot Z)) \ .$$

A natural generalization of this is:

$$(6.2) \qquad \vartheta^{Q}(Z,\Omega) = \sum_{N\in\mathbb{Z}^{(g,h)}} \exp(\pi i\ \mathrm{tr}({}^t N\cdot\Omega\cdot N\cdot Q) + 2\pi i\ \mathrm{tr}({}^t N\cdot Z))$$

where $Q$ is a positive definite rational $h\times h$ matrix and the variable $Z$ lies in $\mathbb{C}^{(g,h)}$ (g h complex matrices), and $\Omega$ lies in $\mathfrak{h}_g$.

$\vartheta^Q$ may be reduced by the old $\vartheta$ if we define a map

$$\begin{gathered} \Omega \longmapsto \Omega \otimes Q \\ \mathfrak{h}_g \longrightarrow \mathfrak{h}_{hg} \end{gathered}$$

where $\Omega\otimes Q$ is the $gh\times gh$ matrix given by

$$(\Omega\otimes Q)_{ih+j,kh+\ell} = \Omega_{i+1,k+1} Q_{j\ell}, \quad 0 \le i,k < g, \quad 1 \le j,\ell \le h.$$

If we rewrite $Z$ as a column:

$$\mathrm{vec}(Z)_{ih+j} = Z_{i+1,j} \qquad 0 \le i < g,\ 1 \le j \le h$$

then it is immediate that

$$\vartheta^{Q}(Z,\Omega) = \vartheta(\mathrm{vec}(Z),\ \Omega\otimes Q).$$

Under the same map, the theta functions with characteristics give us:

$$\vartheta^Q[{}^A_B](Z,\Omega) = \sum_{N\in\mathbb{Z}^{(g,h)}} \exp(\pi i\ \mathrm{tr}({}^t(N+A)\Omega\cdot(N+A)\cdot Q) + 2\pi i\ \mathrm{tr}({}^t(N+A)(Z+B)))$$

$$A,B \in \mathbb{Q}^{(g,h)}.$$

Setting $Z = 0$, one may think of the functions

$$\Omega \longmapsto \vartheta^Q[{}^A_B](0,\Omega)$$

as being a natural basis of the vector space of all functions

$$\vartheta^Q[f](\Omega) = \sum_{N\in\mathbb{Q}^{(g,h)}} f(N)\cdot\exp(\pi i\ \mathrm{tr}({}^tN\Omega NQ))$$

$$f \in \mathcal{S}(\mathbb{Q}^{(g,h)}).$$

Thus all our previous ideas generalize to this setting. In these terms, we may generalize (6.1) as follows:

<u>Theorem</u> (6.3): <u>The functions</u> $\vartheta^Q(Z,\Omega)$ <u>satisfy</u>:

i) <u>if</u> $Q = \left(\begin{array}{c|c} Q_1 & 0 \\ \hline 0 & Q_2\end{array}\right)$, <u>then</u>

$$\vartheta^Q(Z,\Omega) = \vartheta^{Q_1}(Z_1,\Omega)\cdot\vartheta^{Q_2}(Z_2,\Omega)$$

<u>where</u> $Z = (Z_1,Z_2)$.

ii) <u>if</u> $Q' = {}^tT\cdot Q\cdot T$ <u>where</u> $T \in \mathbb{Q}^{(h,h)}$, $Q',Q$ <u>both positive definite</u> $h\times h$ <u>rational symmetric matrices</u>

$$(R^{T,Q})\qquad \vartheta^{Q'}(Z\cdot T,\Omega) = d^{-1}\cdot\sum_{A\in K_1,B\in K_2} \exp(\pi i\ \mathrm{tr}({}^tA\Omega AQ+2{}^tA\cdot(Z+B)))\cdot\vartheta^Q(Z+\ell AQ+B,\Omega)$$

where

$$K_1 = \mathbb{Z}^{(g,h)}.{}^tT/\mathbb{Z}^{(g,h)}.{}^tT \cap \mathbb{Z}^{(g,h)}$$

$$K_2 = \mathbb{Z}^{(g,h)}.T^{-1}/\mathbb{Z}^{(g.h)}.T^{-1} \cap \mathbb{Z}^{(g,h)}$$

$$d = \# K_2 .$$

In particular, if $Q = Q' = I_h$, $(R^{\sigma,Q})$ reduces to $(R^T)$. The proof is exactly the same as that of (6.1). A clearer way to state (6.3), perhaps, is via the functions $\vartheta^Q[f](\Omega)$. Note that to prove (6.3) for all Z, it certainly suffices to prove it for $Z = \Omega AQ+B$, $A,B \in \mathbb{Q}^{(g,h)}$, hence $(R^{T,Q})$ reduces to proving corresponding identity

$$(R^{T,Q}_{ch}) \qquad \vartheta^{Q'}\begin{bmatrix} A.{}^tT^{-1} \\ B\cdot T \end{bmatrix}(Z\cdot T,\Omega) = d^{-1} \sum_{A'\in K_1, B'\in K_2} \exp(-2\pi i \operatorname{tr} {}^tB'.A).\vartheta^Q\begin{bmatrix} A+A' \\ B+B' \end{bmatrix}(Z,\Omega)$$

for all A,B but with $Z = 0$.

When $Z = 0$, it is simply the explicit form , in terms of standard bases of $\mathcal{S}(\mathbb{Q}^{(g,h)})$ of the formula:

$$(R^{T,Q}_{nat}) \qquad \vartheta^{Q'}[f'](\Omega) = \vartheta^Q[f](\Omega)$$

$$\text{where} \quad f'(N) = f(N\cdot {}^tT) .$$

At this point, the proof reduces to the totally obvious calculation:

$$\begin{aligned} \vartheta^Q[f](\Omega) &= \sum_{N\in\mathbb{Q}^{(g,h)}} f(N)\cdot\exp({}^tN\Omega NQ) \\ &= \sum_{N\in\mathbb{Q}^{(g,h)}} f(N)\cdot\exp({}^tN\cdot\Omega\cdot N\cdot {}^tT^{-1}Q'\cdot T^{-1}) \\ &= \sum_{M\in\mathbb{Q}^{(g,h)}} f(M\cdot {}^tT)\cdot\exp({}^tM\cdot\Omega\cdot M\cdot Q') \quad \text{where } M = N\cdot {}^tT^{-1} \\ &= \vartheta^{Q'}[f'](\Omega). \end{aligned}$$

(6.3) has the important Corollary:

Corollary (6.4). i) For all Q and f, $\vartheta^Q[f](\Omega)$ is a modular form in $\Omega$ of weight h/2 and some level.

ii) For each Q

$$f \longmapsto \vartheta^Q[f]$$

is a map

$$w_Q: \mathcal{S}(\mathbb{Q}^{(g,h)}) \longrightarrow \{\text{v. sp. of modular forms of wt. } h/2, \text{ any level}\}$$

called the Weil mapping associated to Q, and the image depends only on isomorphism type of the rational quadratic form ${}^t x.Q.x$ in h variables.

iii) Under multiplication of modular forms, if $Q = \left(\begin{array}{c|c} Q' & 0 \\ \hline 0 & Q'' \end{array}\right)$, then

$$\text{Image}(w_Q) = \text{Image}(w_{Q'}) \cdot \text{Image}(w_{Q''}) \ .$$

Proof: (iii) is a restatement of (6.3.i), and the 2nd half of (ii) is a restatement of $(R^{T,Q}_{nat})$. Now any Q can be diagonalized over $\mathbb{Q}$, and the product $f_1, f_2$ of modular forms of wt. $n_1, n_2$ is a modular form of wt. $n_1+n_2$. So (i) follows from the fact that $\vartheta[f](\ell\Omega)$ is a modular form of weight 1/2 for all $\ell \geq 1$. QED

The 2 fundamental problems in the analysis of theta functions as functions of $\Omega$ are the description of the image and kernel of $w_Q$. For example, one might ask whether

$$\begin{aligned} \text{Ker}(w_Q) = \ & \text{span of the differences } f-f', \\ & \text{where } f'(N) = f(N\cdot{}^tT), \\ & T \in \text{orthogonal gp. for } Q \text{ over } \mathbb{Q} \end{aligned}$$

so that $(R^{T,Q}_{nat})$ gives the full kernel? I don't know if this is true or not. Also, one might ask whether $\mathrm{Im}(w_Q)$ (or $\mathrm{Im}(w_{I_n})$) contains all "cusp" forms if h is bigger than some simple function of g.

As functions of $\vec{z}$ for fixed $\Omega$, however, we saw in §1 how to produce from $\vartheta$ <u>bases</u> for the vector spaces $R^{\Omega}_{\ell}$ of quasi-periodic functions in $\vec{z}$ of each weight. With Riemann's theta relation, we can go further and work out explicitly the multiplication table of the ring $\sum_{\ell} R^{\Omega}_{\ell}$ in terms of these bases. To do this, we apply $(R^{T,Q}_{ch})$ with

$$h = 2$$

$$Q = \begin{pmatrix} n_1+n_2 & 0 \\ 0 & n_1 n_2(n_1+n_2) \end{pmatrix}$$

$$Q' = \begin{pmatrix} n_1 & 0 \\ 0 & n_2 \end{pmatrix}$$

$$T = \frac{1}{(n_1+n_2)}\begin{pmatrix} n_1 & n_2 \\ 1 & -1 \end{pmatrix}$$

$(R^{T,Q}_{ch})$ works out to part (i) of the following Proposition:

<u>Proposition</u> (6.4) i) <u>For all</u> $n_1, n_2 \geq 1$,

$$\vartheta\begin{bmatrix} \vec{a}/n_1 \\ 0 \end{bmatrix}(z_1, n_1\Omega)\cdot\vartheta\begin{bmatrix} \vec{b}/n_2 \\ 0 \end{bmatrix}(z_2, n_2\Omega) = \sum_{\vec{d}\in\mathbb{Z}^g/(n_1+n_2)\mathbb{Z}^g} \vartheta\begin{bmatrix} \frac{n_1\vec{d}+\vec{a}+\vec{b}}{n_1+n_2} \\ 0 \end{bmatrix}(z_1+z_2, (n_1+n_2)\Omega)\cdot$$

$$\cdot\vartheta\begin{bmatrix} \frac{n_1 n_2\vec{d}+n_2\vec{a}-n_1\vec{b}}{n_1 n_2(n_1+n_2)} \\ 0 \end{bmatrix}(n_2 z_1 - n_1 z_2, n_1 n_2(n_1+n_2)\Omega) .$$

ii) <u>For all</u> $n \geq 1$, <u>let</u>

$$f^{(n)}_{\vec{a}}(z) = \vartheta\begin{bmatrix} \vec{a}/n \\ 0 \end{bmatrix}(nz, n\Omega)$$

<u>be the basis of</u> $R^{\Omega}_{n}$ <u>of</u> §1. <u>Then</u>:

$$f^{(n_1)}_{\vec a}\cdot f^{(n_2)}_{\vec b} = \sum_{\vec d\in \mathbb{Z}^g/(n_1+n_2)\mathbb{Z}^g} \vartheta\begin{bmatrix}\frac{n_1n_2\vec d+n_2\vec a-n_1\vec b}{n_1n_2(n_1+n_2)}\\ 0\end{bmatrix}(0,n_1n_2(n_1+n_2)\Omega)\cdot f^{(n_1+n_2)}_{n_1\vec d+\vec a+\vec b}$$

Proof: (ii) follows from (i) by setting $z_1 = n_1z$, $z_2 = n_2z$.

These identities in the case $n_1|n_2$ have been applied by Koizumi (Math. Annalen, 242, 1979, p.127). However the case which has been applied most is when $n_1 = n_2$. In this case, using the simple identity

$$(6.5)\qquad \sum_{\vec e\in \mathbb{Z}^g/n\mathbb{Z}^g} \vartheta\begin{bmatrix}\frac{\vec e+\vec a}{n}\\ 0\end{bmatrix}(n\vec z,\ n^2\Omega) = \vartheta\begin{bmatrix}\vec a\\ 0\end{bmatrix}(\vec z,\Omega)$$

(6.4.i) reduces to the very simple classical:

$$(6.6)\qquad \vartheta\begin{bmatrix}\vec a/n\\ 0\end{bmatrix}(z_1,n\Omega)\cdot\vartheta\begin{bmatrix}\vec b/n\\ 0\end{bmatrix}(z_2,n\Omega) = \sum_{d\in \mathbb{Z}^g/2\mathbb{Z}^g}\vartheta\begin{bmatrix}\frac{\vec d}{2}+\frac{\vec a+\vec b}{2n}\\ 0\end{bmatrix}(z_1+z_2,\ 2n\Omega)\cdot\vartheta\begin{bmatrix}\frac{\vec d}{2}+\frac{\vec a-\vec b}{2n}\\ 0\end{bmatrix}(z_1-z_2,\ 2n\Omega)\ .$$

We give 2 applications of this. In the first, we assume n is even and $n \geq 2$ and, following §1, embed the torus $\mathbb{C}^g/L_\Omega$ into $\mathbb{P}^{N-1}$, $N = n^g$, by:

$$\vec z \longmapsto \left(\cdots,\ \vartheta\begin{bmatrix}\vec a/n\\ 0\end{bmatrix}(n\vec z,n\Omega),\cdots\right)_{\vec a\in \mathbb{Z}^g/n\mathbb{Z}^g}\ .$$

Then (6.6) gives us a simple set of quadratic equations in $\mathbb{P}^{N-1}$ which vanish on the image. Let $n = 2m$. Substitute $\vec a+m\vec e$ for $\vec a$, $\vec b-m\vec e$ for $\vec b$ in (6.6), multiply by $\exp(\pi i\,{}^t\vec e.\vec f)$ and sum over $\vec e \in \mathbb{Z}^g/2\mathbb{Z}^g$. This gives:

$$\sum_{\vec{e}\in\mathbb{Z}^g/2\mathbb{Z}^g} \exp(\pi i\, {}^t\vec{e}\cdot\vec{f})\,\vartheta\begin{bmatrix}\frac{\vec{a}}{n}+\frac{\vec{e}}{2}\\ 0\end{bmatrix}(z_1,n\Omega)\cdot\vartheta\begin{bmatrix}\frac{\vec{b}}{n}-\frac{\vec{e}}{2}\\ 0\end{bmatrix}(z_2,n\Omega)$$

$$=\left[\sum_{\vec{d}\in\mathbb{Z}^g/2\mathbb{Z}^g} \exp(\pi i\, {}^t\vec{d}\cdot\vec{f})\,\vartheta\begin{bmatrix}\frac{\vec{d}}{2}+\frac{a+b}{2n}\\ 0\end{bmatrix}(z_1+z_2,2n\Omega)\right]\cdot\left[\sum_{\vec{d}\in\mathbb{Z}^g/2\mathbb{Z}^g}' \exp(\pi i\, {}^t d\cdot\vec{f})\,\vartheta\begin{bmatrix}\frac{\vec{d}}{2}+\frac{a-b}{2n}\\ 0\end{bmatrix}(z_1-z_2,2n\Omega)\right]$$

Setting $z_1 = z_2 = nz$, and writing

$$f_{\vec{a}}^{(n)}(z) = \vartheta\begin{bmatrix}\vec{a}/n\\ 0\end{bmatrix}(nz,n\Omega), \quad \langle\vec{e},\vec{f}\rangle = \exp(\pi i\, {}^t\vec{e}\cdot\vec{f})$$

we find the identities

$$\lambda_1\cdot\sum_{\vec{e}\in\mathbb{Z}^g/2\mathbb{Z}^g} \langle\vec{e},\vec{f}\rangle\, f^{(n)}_{\vec{a}+m\vec{e}}(z)\cdot f^{(n)}_{\vec{b}+m\vec{e}}(z)$$

$$= \lambda_2\cdot\sum_{e\in\mathbb{Z}^g/2\mathbb{Z}^g} \langle\vec{e},\vec{f}\rangle\cdot f^{(n)}_{\vec{c}+m\vec{e}}(z)\cdot f^{(n)}_{\vec{d}+m\vec{e}}(z)$$

whenever

$$\vec{a}+\vec{b} \equiv \vec{c}+\vec{d} \mod n\mathbb{Z}^g$$

$$\vec{f} \in \mathbb{Z}^g/2\mathbb{Z}^g$$

where the constants are given by:

$$\lambda_1 = \sum_{\vec{e}\in\mathbb{Z}^g/2\mathbb{Z}^g} \langle\vec{e},\vec{f}\rangle\cdot f^{(2n)}_{\vec{c}-\vec{d}+n\vec{e}}(0)$$

$$\lambda_2 = \sum_{\vec{e}\in\mathbb{Z}^g/2\mathbb{Z}^g} \langle\vec{e},\vec{f}\rangle\cdot f^{(2n)}_{\vec{a}-\vec{b}+n\vec{e}}(0)$$

This means that if the homogeneous coordinates in $\mathbb{P}^{N-1}$ are labelled $X_{\vec{a}}$, $\vec{a} \in \mathbb{Z}^g/n\mathbb{Z}^g$, then the tori $\mathbb{C}^g/L_\Omega$ in $\mathbb{P}^{N-1}$ satisfies

$$(6.7)\qquad \lambda_1 \sum \langle \vec{e},\vec{f}\rangle\, X_{\vec{a}+m\vec{e}} \cdot X_{\vec{b}+m\vec{e}} = \lambda_2 \sum \langle \vec{e},\vec{f}\rangle\, X_{\vec{c}+m\vec{e}} \cdot X_{\vec{d}+m\vec{e}}$$

for all $\vec{a}+\vec{b} \equiv \vec{c}+\vec{d}$, $\vec{f} \in \mathbb{Z}^g/2\mathbb{Z}^g$. ($\lambda_1, \lambda_2$ as above). In fact, it is proven in Mumford, Inv. Math., vol. 1, 1966, pp. 341-349, that <u>these quadratic equations are a complete set of equations for the image of the torus.</u>*

As a second application, in (6.6) take $n = 2^k$, $z_1 = z_2 = 2^k z$ and consider the bases

$$f_{\vec{a}}^{(2^k)}(z) = \vartheta\begin{bmatrix}\vec{a}/2k\\0\end{bmatrix}(2^k z, 2^k\Omega)$$

of $R_{2^k}^{(\Omega)}$. In terms of these bases, we get a very simple and beautiful <u>multiplication table</u>

$$(6.8)\qquad f_{\vec{a}}^{(2^k)} \cdot f_{\vec{b}}^{(2^k)} = \sum_{\substack{\vec{c},\vec{d} \in \mathbb{Z}^g/2^{k+1}\mathbb{Z}^g \\ \vec{c}+\vec{d} \equiv 2\vec{a} \\ \vec{c}-\vec{d} \equiv 2\vec{b}}} f_{\vec{c}}^{(2^{k+1})} \cdot f_{\vec{d}}^{(2^{k+1})}(0).$$

This implies the identity

$$(6.9.a)\qquad f_{\vec{a}}^{(2^k)}(0) \cdot f_{\vec{b}}^{(2^k)}(0) = \sum_{\substack{\vec{c},\vec{d} \in \mathbb{Z}^g/2^{k+1}\mathbb{Z}^g \\ \vec{c}+\vec{d} \equiv 2\vec{a} \\ \vec{c}-\vec{d} \equiv 2\vec{b}}} f_{\vec{c}}^{(2^{k+1})}(0) \cdot f_{\vec{d}}^{(2^{k+1})}(0)$$

on the modular forms $f_{\vec{a}}^{(2^k)}(0) = \vartheta\left[\begin{smallmatrix}\vec{a}/2^k\\0\end{smallmatrix}\right](0, 2^k\Omega)$. In fact, <u>any solution of these identities plus the further identities:</u>

$$(6.9.b)\qquad f_{\vec{a}}^{(2^k)}(0) = \sum_{\substack{\vec{b} \in \mathbb{Z}^g/2^{k+2}\mathbb{Z}^g \\ \vec{b} \equiv 2\vec{a} \bmod 2^{k+1}\mathbb{Z}^g}} f_{\vec{b}}^{(2^{k+2})}(0)$$

(a special case of (6.5))

*More precisely, the ideal they generate equals the full ideal of $\mathbb{C}^g/L$ in sufficiently large degrees.

(6.9c) $$f_{\vec{a}}^{(2^k)}(0) = f_{-\vec{a}}^{(2^k)}(0)$$

comes from some $\Omega$ or a "limit" of $\Omega$'s. This is proven in Mumford, Inv. Math., 3, 1967, §10-11 , where a complete description of the "limiting" values of the $f_{\vec{a}}^{(2^k)}(0)$ is also given. In terms of inverse limits, as in Ch. I, §17, we can restate the result as:

(6.10)
$$\varprojlim_k \left[\mathfrak{h}_g/\Gamma_{2^k} \cup \begin{matrix}\text{certain}\\ \text{cusps}\end{matrix}\right] \cong$$

$$\operatorname{Proj}\left[\mathbb{C}[\ldots, Y_{k,\vec{a}}, \ldots]_{\substack{\text{all } k\geq 1\\ \vec{a}\in\mathbb{Z}^g/2^k\mathbb{Z}^g}} \Big/ \begin{matrix}\text{identities 6.9.a,6.9.b,6.9.c}\\ \text{with } f_{\vec{a}}^{(2^k)}(0) \text{ replaced}\\ \text{by } Y_{k,\vec{a}}\end{matrix}\right]$$

There is another interpretation of the "data" $\{f_{\vec{a}}^{(2^k)}(0)\}$ and the identities 6.9.a,6.9.b,6.9.c which is quite beautiful. Note that

$$f_{\vec{a}}^{(2^{2k})}(0) = \vartheta\begin{bmatrix}\vec{a}/2^{2k}\\ 0\end{bmatrix}(0, 2^{2k}\Omega)$$

$$= \sum_{\vec{n}\in\left(2^k\mathbb{Z}^g + \frac{\vec{a}}{2^k}\right)} \exp(\pi i\ {}^t\vec{n}\cdot\Omega\cdot\vec{n})$$

while

$$f_{\vec{a}}^{(2^{2k+1})}(0) = \sum_{\vec{n}\in\left(2^k\mathbb{Z}^g + \frac{\vec{a}}{2^{k+1}}\right)} \exp(\pi i\ {}^t\vec{n}\cdot 2\Omega\cdot\vec{n}) .$$

Equivalently, we may define <u>2 measures</u> $\mu,\nu$ <u>on</u> $\mathbb{Q}_2^g$ by:

(6.11)
$$\mu(U) = \sum_{\vec{n} \in U \cap \mathbb{Z}[\frac{1}{2}]^g} \exp(\pi i \, {}^t\vec{n} \cdot \Omega \cdot \vec{n})$$
$$\nu(U) = \sum_{\vec{n} \in U \cap \mathbb{Z}[\frac{1}{2}]^g} \exp(\pi i \, {}^t\vec{n} \cdot 2\Omega \cdot \vec{n})$$

for all open sets $U \subset \mathbb{Q}_2^g$. Then

$$\mu(2^k \mathbb{Z}_2^g + \frac{\vec{a}}{2^k}) = f_{\vec{a}}^{(2^{2k})}(0)$$
$$\nu(2^k \mathbb{Z}_2^g + \frac{\vec{a}}{2^{k+1}}) = f_{\vec{a}}^{(2^{2k+1})}(0).$$

6.9.b is subsumed under the fact that $\mu$ and $\nu$ are measures, and 6.9.c says they are even measures:

$$\mu(-U) = \mu(U); \quad \nu(-U) = \nu(U).$$

A little calculation will show that 6.9.a says that $\mu$ and $\nu$ are linked as follows:

(6.12) let $\xi : \mathbb{Q}_2^g \times \mathbb{Q}_2^g \longrightarrow \mathbb{Q}_2^g \times \mathbb{Q}_2^g$ be
$$\xi(\vec{x}, \vec{y}) = (\vec{x}+\vec{y}, \vec{x}-\vec{y}).$$

Then
$$\xi_*(\mu \times \mu) = \nu \times \nu.$$

Thus (6.10) can be restated as:

(6.13)
$$\varprojlim \left[ \mathfrak{H}_g / \Gamma_{2k} \cup \begin{matrix}\text{certain}\\ \text{cusps}\end{matrix} \right] \cong \left\{ \begin{matrix} \text{pairs } \mu, \nu \text{ of even measures on } \mathbb{Q}_2^g \\ \text{saitsfying 6.12 mod scalars} \end{matrix} \right\}$$

(See Mumford, Inv. Math. 3, p. 116 ).

## §7. Theta functions with harmonic coefficients.

Starting with the functional equation for $\vartheta(\vec{z},\Omega)$, we have seen in §5 that we can define a large space of modular forms of weight 1/2 by

$$\vartheta[f](\ell\Omega) = \sum f(\vec{n})\exp(\pi i\ell\, {}^t\vec{n}\cdot\Omega\cdot\vec{n}), \qquad f \in \mathcal{S}(\mathbb{Q}^g).$$

The functions $\vartheta[f](\ell\Omega)$ are all linear combinations with elementary exponential factors of the functions

$$\vartheta(\Omega\vec{a}+\vec{b},\ell\Omega), \qquad \vec{a},\vec{b} \in \mathbb{Q}^g$$

obtained by restricting $\vec{z}$ to a point of finite order mod $L_\Omega$. We ask: are there other ways of getting modular forms from $\vartheta(\vec{z},\Omega)$? In fact, another way is by differentiating $\vartheta$ with respect to $\vec{z}$ and then setting $\vec{z} = 0$, or more generally $\vec{z} = \Omega\vec{a}+\vec{b}$. To illustrate this, look again at the one-variable case:

$$\frac{\partial}{\partial z}\vartheta\left[\begin{smallmatrix}a\\b\end{smallmatrix}\right](z,\tau)\Big|_{z=0} = \sum_{n\in\mathbb{Z}} \frac{\partial}{\partial z}\exp\left(\pi i(n+a)^2\tau + 2\pi i(n+a)(z+b)\right)\Big|_{z=0}$$

$$= 2\pi i \circ \sum_{n\in\mathbb{Z}+a} n\cdot\exp(\pi i n^2\tau+2\pi i nb) .$$

If we differentiate the functional equation

$$\vartheta\left[\begin{smallmatrix}a\\b\end{smallmatrix}\right]\left(\frac{z}{\gamma\tau+\delta}, \frac{\alpha\tau+\beta}{\gamma\tau+\delta}\right) = \zeta\cdot\sqrt{\gamma\tau+\delta}\cdot e^{\frac{\pi i\gamma z^2}{\gamma\tau+\delta}}\cdot\vartheta\left[\begin{smallmatrix}a\\b\end{smallmatrix}\right](z,\tau)$$

(where $\begin{pmatrix}\alpha & \beta\\ \gamma & \delta\end{pmatrix} \in SL(2,\mathbb{Z})$ is in a small enough congruence subgroup), with respect to z, and set z = 0, we see that

$$\frac{\partial\vartheta[{a \atop b}]}{\partial z}\left(0, \frac{\alpha\tau+\beta}{\gamma z+\delta}\right) = \zeta(\sqrt{\gamma\tau+\delta})^{3/2} \cdot \frac{\partial\vartheta[{a \atop b}]}{\partial z}(0,\tau),$$

i.e., $\partial\vartheta[{a \atop b}]/\partial z$ is a modular form of weight 3/2. But if we differentiate the functional equation twice, we see, for instance, that

$$\frac{\partial^2\vartheta[{a \atop b}]}{\partial z^2}(0,\tau) = -4\pi^2 \sum_{n\in\mathbb{Z}+a} n^2\exp(\pi i n^2\tau+2\pi i n b)$$

is not a modular form. But persevere! A longer calculation will show you that

$$\frac{1}{3}\cdot\frac{\partial^3\vartheta[{a \atop b}]}{\partial z^2}(0,\tau)\cdot\vartheta[{a' \atop b'}](0,\tau) - \frac{\partial\vartheta[{a \atop b}]}{\partial z}(0,\tau)\cdot\frac{\partial\vartheta^2[{a' \atop b'}]}{\partial z^2}(0,\tau)$$

is again a modular form, now of weight 4. The point is that the functional equation introduces a factor $e^{\lambda z^2}$ and we need to form combinations of the z-derivatives at z = 0 which are invariant under substitutions $\vartheta(z) \longmapsto e^{\lambda z^2}\cdot\vartheta(z)$. Written out, this last modular form is

$$(2\pi i)^4\cdot \sum_{\substack{n\in\mathbb{Z}+a\\ m\in\mathbb{Z}+a'}} P(n,m)e^{\pi i(n^2+m^2)\tau}\cdot e^{2\pi i nb+mb'}$$

or (in terms of theta series for the quadratic form $X^2+Y^2$):

$$P\left(\frac{\partial}{\partial z_1}, \frac{\partial}{\partial z_2}\right)\vartheta^{I_2}\begin{bmatrix} a & a' \\ b & b'\end{bmatrix}((z_1,z_2),\tau)\Big|_{z_1=z_2=0}$$

where

$$P(n,m) = \frac{1}{3}n^3 - nm^2 .$$

Note that P is a spherical harmonic polynomial. For several variables, the situation is of course more complicated, as we have g partials $\partial/\partial z_i$. In fact, the natural thing to expect to find are vector-valued modular forms.

Here is what happens:

Definition (7.1): Let $\tau: GL(g,\mathbb{C}) \longrightarrow GL(N,\mathbb{C})$ be an N-dimensional polynomial representation of $GL(g,\mathbb{C})$ or a 2-valued representation given by $\tau(A) = \tau_0(A)\cdot\sqrt{\det A}$ where $\tau_0$ is a polynomial representation. Then an N-tuple $\vec{f} = (f_1,\cdots,f_N)$ of holomorphic functions of $\Omega$ is called a vector-valued modular form of level $\Gamma$, type $\tau$ if

$$f_i((A\Omega+B)(C\Omega+D)^{-1}) = \sum_{j=1}^{n} \tau(C\Omega+D)_{ij} f_j(\Omega)$$

for all $1 \leq i \leq n$, $\begin{pmatrix} A & B \\ C & D \end{pmatrix} \in \Gamma$.

Definition (7.2): Let $X \in \mathbb{C}^{(g,h)}$ be a matrix variable. A polynomial $P(X)$ is called pluri-harmonic if

$$\sum_{k=1}^{h} \frac{\partial^2 P}{\partial X_{ik}\,\partial X_{jk}} \equiv 0, \qquad 1 \leq i,j \leq g.$$

We shall denote by $\mathbb{H}_\ell$ the vector space of all pluri-harmonic polynomials P which are homogeneous of degree $\ell$. Note that if $P(X)$ is pluri-harmonic, $P'(X) = P(A\cdot X\cdot B)$ is also pluri-harmonic for all $A \in GL(g,\mathbb{C})$, $B \in O(h,\mathbb{C})$. Thus $GL(g,\mathbb{C}) \times O(h,\mathbb{C})$ acts on $\mathbb{H}_\ell$.

Definition (7.3): For all

$$P \in \mathbb{H}_\ell,\quad Q \text{ rational pos. def. } h\times h \text{ symmetric},\quad f \in \mathcal{S}(\mathbb{Q}^{(g,h)}),$$

let

$$\vartheta^{P,Q}[f](\Omega) = \sum_{N\in\mathbb{Q}^{(g,h)}} f(N)\cdot P(N\cdot\sqrt{Q})\cdot\exp\left(\pi i\ \mathrm{tr}({}^tN\cdot\Omega\cdot N\cdot Q)\right).$$

The main result is this:

Theorem 7.4: Let $V \subset \mathbb{H}_\ell$ be a subspace invariant under $GL(g,\mathbb{C})$, let $\{P_\alpha\}$ be a basis of V and let $GL(g,\mathbb{C})$ act on V via the representation $\tau$:

$$P_\alpha(A\cdot X) = \sum_\beta \tau_{\alpha\beta}(A)\cdot P_\beta(X).$$

Then for all Q,f, the sequence of functions $\left\{\vartheta^{P_\alpha,Q}[f]\right\}$ is a vector-valued modular form of type $\tau\otimes\det^{h/2}$ and suitable $\Gamma$.

A word about the history of this result: Hecke, Maass and others have investigated various types of theta series with harmonic coefficients and proved this Theorem in many cases. Kashiwara-Vergne (Inv. Math., 44 (1978)) worked out very completely these results from a representation-theoretic point of view and also decomposed $\mathbb{H}_\ell$ as a representation of $GL(g,\mathbb{C})\times O(h,\mathbb{C})$. Theorem (7.4) in its full generality was proven independently by Freitag (Math. Annalen, 254 (1980), pp. 27-51) and T. Oda (Theta series of definite quad. forms, to appear). The approach that we use is based on the ideas of Barsotti (Considerazioni sulle funzioni theta, Symp. Math., 3 (1970), p. 247) analyzing theta functions from an algebro-geometric point of view. We will describe both Kashiwara-Vergne's results and Barsotti's in more detail in Ch. IV and consider only the purely classical-analytic results in this Chapter.

The theorem could be proven using generators for $\Gamma_{1,2}$ and allowing a transformation on Q,f, following the ideas of §5. However we can also, following Barsotti, draw a proof directly out of the ideas of the examples above, i.e., by differentiating the functional equation for $\vartheta^Q(Z,\Omega)$ with respect to Z. To do this we need first to see clearly why pluri-harmonic polynomials come in.

Put an inner product on the polynomial ring $\mathbb{C}[\cdots,Z_{ij},\cdots]$, $1 \le i \le g$, $1 \le j \le h$, by

$$\langle P,Q\rangle = (P(\cdots,\partial/\partial Z_{ij},\cdots)\bar{Q})(0).$$

Note that 2 monomials $Z^\alpha, Z^\beta$ are perpendicular if $\alpha \neq \beta$ and $\langle Z^\alpha, Z^\beta\rangle$ is a positive integer, hence $\langle , \rangle$ is positive definite Hermitian. Let

$$\mathfrak{N} = \left\{\begin{matrix}\text{ideal in } \mathbb{C}[\cdots, Z_{ij}, \cdots] \text{ generated by} \\ W_{ij} = \sum_{k=1}^{h} Z_{ik} Z_{jk} \quad .\end{matrix}\right\}$$

Then we have:

Proposition 7.5. i) $\mathbb{C}[\cdots, Z_{ij}, \cdots] \cong \mathbb{H} \oplus \mathfrak{N}$ and $\mathbb{H}, \mathfrak{N}$ are perpendicular with respect to the above inner product,

ii) Let $\mathcal{O}$ = ring of analytic functions on $\mathbb{C}^{(g,h)}$ near 0, and for all $P \in \mathbb{C}[\cdots, Z_{ij} \cdots]$, define

$$\delta_P : \mathcal{O} \longrightarrow \mathbb{C}$$

by $\delta_P(f) = (P(\cdot\cdot, \partial/\partial Z_{ij}, \cdot\cdot)f)(0)$.

Then $P$ is pluri-harmonic iff

$$\delta_P(f) = \delta_P(e^{\operatorname{tr} {}^t Z \cdot C \cdot Z} \cdot f), \text{ all symmetric } h \times h C. \tag{7.6}$$

Proof: To prove (i), note that

$$\begin{aligned} P \in \mathbb{H}_k \iff & \sum_k \frac{\partial^2}{\partial Z_{ik} \partial Z_{jk}} P \equiv 0 \\ \iff & \left((R(\frac{\partial}{\partial Z_{ij}}) \circ \sum \frac{\partial^2}{\partial Z_{ik} \partial Z_{jk}}) P\right)(0) = 0, \quad \text{all } R \\ \iff & (R(\frac{\partial}{\partial Z_{ij}}) P)(0) = 0, \quad \text{all } R \in \mathfrak{N} \\ \iff & P \in \mathfrak{N}^{\perp}. \end{aligned}$$

To prove (ii), note that it suffices to take f to be a polynomial because $\delta_P(f)$ depends only on $f \in \mathcal{O}/\mathfrak{m}^N$, some N, and polynomials map onto $\mathcal{O}/\mathfrak{m}^N$. Moreover, we are asking invariance w.r.t. a grup so

$$\delta_P \text{ has invariance (7.6)} \iff \frac{\partial}{\partial c_{pq}} \delta_P(e^{tr\,{}^tZ\cdot C\cdot Z}\cdot f)\Big|_{C=0} = 0, \text{ all } f,p,q$$

$$\iff P\left(\frac{\partial}{\partial z_{ij}}\right)\left(\sum_{k=1}^{h} z_{pk}z_{qk}\cdot f\right)(0) = 0, \quad \text{all } f,p,q$$

$$\iff P\left(\frac{\partial}{\partial z_{ij}}\right) f\,(0) = 0, \qquad \text{all } f \in \mathcal{N}$$

$$\iff P \in \mathcal{N}^{\perp}$$

$$\iff P \in \mathbb{H}.$$

QED

Corollary (7.7): If P is pluri-harmonic and Q is a real positive definite $h\times h$ symmetric matrix, then $P^*(X) = P(X\cdot\sqrt{Q})$ satisfies

$$\delta_{P^*}(f) = \delta_{P^*}(e^{tr\,{}^tZ\cdot C\cdot Z\cdot Q^{-1}}f), \text{ all } C.$$

Proof: Substitute $Z = W\cdot \sqrt{Q}^{-1}$ in (7.6).

To prove the Theorem, note first that the span of the $\vartheta^{P,Q}[f]$'s for fixed Q, depends only on the rational equivalence class of Q because

$$\vartheta^{P',Q'}[f'] = \vartheta^{P,Q}[f]$$

if $Q' = {}^tAQA$, $f'(N) = f(N\cdot{}^tA)$, $P'(X) = P(X(\sqrt{Q'}^{-1}\cdot{}^tA\cdot\sqrt{Q}))$.

Here $A \in GL(h,\mathbb{Q})$ (and note that $\sqrt{Q'}^{-1} \cdot {}^tA \cdot \sqrt{Q}$ is orthogonal so that $P'$ is again pluri-harmonic). Therefore we can assume $Q_{ij} = \ell_i \delta_{ij}$ and we may also assume $f = f_1 \otimes \cdots \otimes f_h$, $f_i \in \mathcal{S}(\mathbb{Q}^g)$. Then

$$(2\pi i)^h \cdot \vartheta^{P_\alpha, Q}[f](\Omega) = \left[ P_\alpha^*\left(\cdots, \frac{\partial}{\partial z_{ij}}, \cdots\right) \left( \prod_{i=1}^{h} \vartheta[f_i](\vec{z}_i, \ell_i \Omega) \right) \right]_{Z=0}$$

where $P_\alpha^*(X) = P_\alpha(X \cdot \sqrt{Q})$, $\vec{z}_i = (z_{1i}, \cdots, z_{gi})$. On the other hand, the functional equation for Riemann's theta function tells us that

$$\prod \vartheta[f_i]\left({}^t(C\Omega+D)^{-1} \cdot \vec{z}_i, \ell_i (A\Omega+B)(C\Omega+D)^{-1}\right) =$$

$$[\pm \det(C\Omega+D)]^{h/2} \exp\left(\pi i \sum_{i=1}^{h} {}^t\vec{z}_i \cdot (C\Omega+D)^{-1} C \cdot \vec{z}_i \cdot \ell_i^{-1}\right) \cdot \prod \vartheta[f_i](\vec{z}_i, \ell_i \Omega)$$

for $\begin{pmatrix} A & B \\ C & D \end{pmatrix}$ in a suitable $\Gamma$. Apply $P_\alpha^*(\cdot\cdot, \partial/\partial z_{ij}, \cdot\cdot)$ to this identity and set $\vec{z}_i = 0$, all $i$. By (7.7) the LHS gives us

$$(\pm \det(C\Omega+D))^{h/2} \cdot \left(P_\alpha^*\left(\cdot\cdot, \frac{\partial}{\partial z_{ij}}, \cdot\cdot\right) \prod \vartheta[f_i](\vec{z}_i, \ell_i \Omega)\right)\Big|_{Z=0}$$

while the RHS gives us

$$\sum \tau((C\Omega+D)^{-1})_{\alpha\beta} \left(P_\beta^*\left(\cdot\cdot, \frac{\partial}{\partial z_{ij}}, \cdot\cdot\right) \prod \vartheta[f_i](\vec{z}_i, \ell_i \Omega)\right)\Big|_{Z=0} .$$

Combining these, we get the function equation for $\{\vartheta^{P_\alpha, Q}[f]\}$.

At this point, we have produced a great quantity of new modular forms, even new scalar modular forms. The most important outstanding problem is to find identities among them. The only non-trivial example is Jacobi's identity (Ch. I, §13) for $g = 1$, and its generalizations to higher $g$ (Fay, Nachr. der Akad. Gottingen, 1979, N$^{\underline{o}}$ 5 ; Igusa, On Jacobi's derivative formula, to appear). Even for $g = 1$, there must be many further identities (e.g., because many modular forms can be represented as theta series in many ways with different P,Q's: cf. Waldspurger, Inv. Math., 50 (1978), p. 135). Is there a systematic way of deriving these from Riemann's theta formula?

In another direction, one of the most interesting applications of these vector-valued theta modular forms is to construct holomorphic differential forms on the Siegel modular variety $\mathfrak{H}_g/Sp(2g,\mathbb{Z})$ (more precisely, on a smooth compactified version of it). This idea is due to Freitag (Math. Ann. 216 (1975), p. 155; Crelle 296 (1977), p. 162) and has been developed by Anderson (Princeton Ph.D. thesis, 1981) and Stillman (Harvard Ph.D. thesis, 1983). We refer the reader to their papers for more details.

MIX
Papier aus verantwortungsvollen Quellen
Paper from responsible sources
FSC® C105338

If you have any concerns about our products,
you can contact us on
**ProductSafety@springernature.com**

In case Publisher is established outside the EU,
the EU authorized representative is:
**Springer Nature Customer Service Center GmbH**
**Europaplatz 3, 69115 Heidelberg, Germany**

Printed by Libri Plureos GmbH
in Hamburg, Germany